ANTENNA THEORY AND APPLICATIONS

ANTENNA THEORY AND APPLICATIONS

Hubregt J. Visser

Holst Centre/imec, The Netherlands

A John Wiley & Sons, Ltd., Publication

This edition first published 2012
© 2012 John Wiley & Sons Ltd

Registered office
John Wiley & Sons Ltd, The Atrium, Southern Gate, Chichester, West Sussex, PO19 8SQ, United Kingdom

For details of our global editorial offices, for customer services and for information about how to apply for permission to reuse the copyright material in this book please see our website at www.wiley.com.

Library of Congress Cataloging-in-Publication Data

Visser, Hubregt J.
  Antenna theory and applications / Hubregt Visser.
      p. cm.
  Includes bibliographical references and index.
  ISBN 978-1-119-99025-3 (cloth)
  1.   Antennas (Electronics)   I.   Title.
  TK7871.6.V55 2012
  621.382′4–dc23

                              2011042247

A catalogue record for this book is available from the British Library.

ISBN: 9781119990253

Typeset in 10/12pt Times by Laserwords Private Limited, Chennai, India
Printed and bound in Singapore by Markono Print Media Pte Ltd

Contents

Preface

This book is derived from a 24-contact-hour, elective course in antenna theory given at Eindhoven University of Technology, The Netherlands. The course is intended for fourth-year students, having a BSc degree in electrical engineering. The students are presumed to have a knowledge of electromagnetic theory and vector analysis.

The original intention in writing this book was to provide a compact, English-language text dealing with the basics of antennas that can be taught in a limited and possibly further shrinking time span. It would have been an alternative to many of the antenna textbooks around that provide too much material for this course. Upon completing the first manuscript, it appeared to be just over one hundred pages, which is too short to put into print. Therefore it was decided to complement the text with examples and design studies of antennas using a commercially distributed full-wave analysis software suite. The choice made was the CST Microwave Studio®, the reason being the familiarity of the author with this software suite. The examples and design studies are, however, described in such a way that any other full-wave analysis software suite that the reader has access to may be used instead. In the appropriate chapters, the theory derived will be used to assess the dimensions of an initial, realistic (i.e. non-ideal) antenna, which will be fine-tuned to the desired characteristics using the full-wave analysis software suite.

The theoretical parts of the book may still be taught in a course of 20–24 contact hours, while the examples and design studies may be left to the student for self-study, or they can be incorporated in a longer course. The CST Microwave Studio® model files are available at a companion website.

The book is organized as follows:

Chapter 1: Introduction. In this chapter a brief history of antennas is presented. The source of electromagnetic radiation is discussed and the mechanism by which radiated fields emerge from an antenna is explained. A brief overview of the antenna types discussed in this book is presented. This chapter is partly taken from [1].

Chapter 2: Antenna System-Level Performance Parameters. Before the theoretical treatment of antennas starts, it is good to have a knowledge of what parameters are important to characterize an antenna and what these parameters mean. This chapter treats this topic in detail since it is considered to be of paramount importance in understanding the 'why' of the mathematics to come. This chapter is taken from [1].

Chapter 3: Vector Analysis. Finally, before beginning with the actual treatment of antennas, experience has shown that it is wise to give a brief 'refresher course' in vector algebra. In this chapter we look at working with the *grad*, *div* and *curl* and introduce the ∇-operator.

Chapter 4: Radiated Fields. In this chapter the calculation of far-fields from general current distributions is introduced and the reciprocity concept is discussed.

Chapter 5: Dipole Antennas. The concepts developed in Chapter 4 are applied here to elementary and finite length dipole antennas.

Chapter 6: Loop Antennas. Here the loop antenna are discussed, the infinitesimal loop antenna is analyzed and the similarities with the infinitesimal dipole antenna are dealt with. As an example, a small printed loop antenna, matched to 50 Ω, is designed.

Chapter 7: Aperture Antennas. In this chapter a general procedure for analyzing aperture antennas is discussed. The theory will be applied to both a horn antenna and a parabolic reflector antenna. As a special case of an aperture antenna, the rectangular microstrip patch antenna will be introduced.

Chapter 8: Array Antennas. In the final chapter the topic of array antennas is explored, but limited to linear array antennas. This chapter is partly taken from [1].

All chapters are concluded with a section of problems. The worked answers are available at a companion website – www.wiley.com/go/visser_antennas.

Reference

1. Hubregt J. Visser, *Array and Phased Array Antenna Basics*, John Wiley & Sons, Chichester, UK, 2005.

Hubregt J. Visser
Veldhoven, The Netherlands

Acknowledgements

I am indebted to the students of 2010 and 2011 taking the course in 'EM antennas and radiation' at the Faculty of Electrical Engineering of Eindhoven University of Technology for proofreading parts of the manuscript and pointing out several typos. In particular I would like to thank Peter Rademakers, Nadejda Roubtsova, Mohadig Rousstia, Stefan Lem, Rob Spijkers and Ziyang Wang.

I am also indebted to the staff of Computer Simulation Technology, especially Frank Demming, Marko Walter and Ulrich Becker, for checking the simulation results and making valuable suggestions.

I thank Tiina Ruonamaa and Susan Barclay from Wiley for their support and – what was unfortunately needed again – incredible patience.

Finally, I would like to thank my wife Dianne and daughters Noa and Lotte for understanding and accepting the many hours of neglect during the writing of this book.

H.J.V.

List of Abbreviations

AC	alternating current
BBC	British broadcasting corporation
CB	Citizens band
CPW	coplanar waveguide
CST	computer simulation technology
EHF	extremely high frequency
EIRP	effective isotropic radiated power
ELF	extremely low frequency
EMC	electromagnetic compatibility
FEM	finite elements method
FIT	finite integration technique
FR	flame retardant
GSM	global system for mobile (communication)
HF	high frequency
HFSS	high frequency structure simulator
IBM	International business machines
IEEE	Institute of electrical and electronics engineers
IFA	inverted F antenna
ILF	infra low frequency
LF	low frequency
LHCP	left-hand circular polarization
MF	medium frequency
MoM	method of moments
MWS	microwave studio
NEC	numerical electromagnetics code
PC	personal computer
PCB	printed circuit board
PEC	perfect electric conductor
PIFA	planar inverted F antenna
Radar	radio detection and ranging
RCS	radar cross section
RF	radio frequency
RHCP	right-hand circular polarization

SHF	super high frequency
SI	system integrity
SLL	side lobe level
SMA	sub miniature A
TEM	transverse electromagnetic
UHF	ultra high frequency
UWB	ultra wideband
VHF	very high frequency

1

Introduction

Antennas have been around now for nearly 125 years. In those 125 years wireless communication has become increasingly important. Personal mobile communication applications are putting huge constraints on the antennas that need to be housed in limited spaces. Therefore the common practice of wireless engineers to consider the antenna as a black-box component is not valid anymore. The modern wireless engineer needs to have a basic understanding of antenna theory. Before we dive into the derivation of antenna characteristics, however, we will – in this chapter – present a brief overview of antenna history and the mechanisms of radiation. Thus, a solid foundation will be presented for understanding antenna characteristics and their derivations.

1.1 The Early History of Antennas

When James Clerk Maxwell, in the 1860s, united electricity and magnetism into electromagnetism, he described light as – and proved it to be – an electromagnetic phenomenon. He predicted the existence of electromagnetic waves at radio frequencies, that is at much lower frequencies than light. In 1886, Maxwell was proven right by Heinrich Rudolf Hertz who – without realizing it himself[1] – created the first ever radio system, consisting of a transmitter and a receiver, see Figure 1.1.

The transmitting antenna, connected to a spark gap at the secondary windings of a conduction coil, was a dipole antenna. The receiving antenna was a loop antenna ending in a second spark gap. Hertz, who conducted his experiments at frequencies around 50 MHz, was able to create electromagnetic waves and to transmit and receive these waves by using antennas. This immediately raises two questions:

1. What is an antenna?
2. How is electromagnetic radiation created?

---

[1] Hertz was not after creating wireless communication but proving the Maxwell equations experimentally.

Antenna Theory and Applications, First Edition. Hubregt J. Visser.
© 2012 John Wiley & Sons, Ltd. Published 2012 by John Wiley & Sons, Ltd.

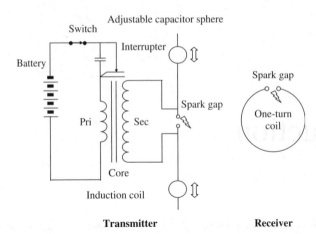

Figure 1.1 Hertz's radio system. With the receiving one-turn loop, small sparks could be observed when the transmitter discharged. From [1].

1.2 Antennas and Electromagnetic Radiation

From the previous it is obvious that

> An antenna is a device for transmitting or receiving electromagnetic waves. An antenna converts electrical currents into electromagnetic waves (transmitting antenna) and vice versa (receiving antenna).

Before we describe this in detail, we will first take a closer look at the origin of electromagnetic radiation.

1.2.1 Electromagnetic Radiation

The source of electromagnetic radiation is accelerated (or decelerated) charge.

Let's start with a static, charged object and have a look at the electric field lines. These lines are the trajectories of a positively charged particle due to this static, charged object. Electric field lines are always directed perpendicular to the surface of a charged object and start and end on charged objects. Electric field lines due to single charged objects start at or extend towards infinity. For a positively charged object, the electric field lines start at the object and extend towards infinity, for a negatively charged object they start at infinity and end at the object.

For explaining the mechanisms of radiation, the direction of the electric field lines does not matter, therefore in Figure 1.2(a), where we show a uniformly moving particle at a certain instant of time, we do not indicate the direction of the field lines.

The uniformly charged particle is accelerated between $t = 0$ and $t = t_1$, see Figure 1.2(b), after which it continues its uniform movement. In Figure 1.2(b) we have indicated the position of the particle at the start ($t = 0$) and at the end ($t = t_1$) of the acceleration.

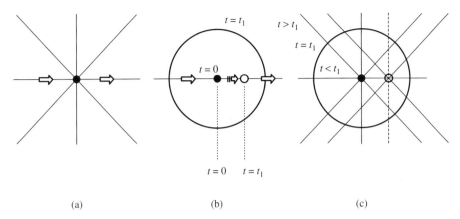

Figure 1.2 Electric field lines of a charged particle. (a) Field lines at a certain moment of time for a uniformly moving charged particle. (b) The particle is accelerated between $t = 0$ and $t = t_1$. The position of an observer, traveling with the speed of light along an electric field line at $t = t_1$ is indicated with the circle. (c) Electric field lines at $t = 0$ and $t = t_1$.

Also indicated is the position of an observer that has moved with the speed of light along a static electric field line from the particle, for the duration of the acceleration (t_1).

In Figure 1.2(c) we repeat Figure 1.2(b), where we now also indicate static electric field lines associated with the particle at $t = 0$ and at $t = t_1$.

We now think of ourselves positioned anywhere on the 'observer circle' and accept that nothing can move faster than the speed of light. Then, everywhere from the 'observer circle' to infinity, the static field lines must follow those associated with the particle position at $t = 0$. Everywhere inside the circle, the static field lines must follow those associated with the particle position at $t = t_1$. Since electric field lines must be continuous, so-called *kinks* must exist at the observer position to make the electric field lines connect, see Figure 1.3.[2]

Having explained the construction of electric field lines for an accelerated charged particle, we can now take a closer look at the electric field lines as a function of time. In Figure 1.4 we look at the electric field lines at different times within the acceleration time interval.

When we take the disturbances, that is the transverse components of the electric field, taken at the subsequent moments and add them in one graph, as in Figure 1.4, we see that these disturbances move out from the accelerated charge at the speed of light. Associated with the changing electric field is a changing magnetic field. Both fields are *in phase*[3] since they are due to a single event. The electric and magnetic fields travel along in phase, their directions being perpendicular to each other. This is what we call an *electromagnetic wave*.

---

[2] The continuous electric field lines are shown a little displaced to clarify the construction from the initial and end position static electric field lines.
[3] As opposed to the situation for a coil or a capacitor.

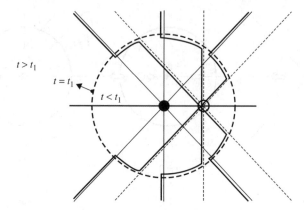

Figure 1.3 The electric field lines of a briefly accelerated charged particle must form *kinks* in order to connect the field lines associated with the initial and end position of the particle, thus forming continuous electric field lines.

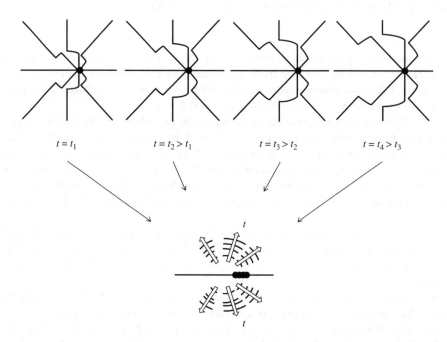

Figure 1.4 The electric field lines of a briefly accelerated charged particle at subsequent instances of time, and the resulting transverse field moving out at the speed of light.

Accelerating (or decelerating) charges may be found in electrically conducting wires at positions were the wire is bent, curved, discontinuous or terminated. Before we discuss the radiation from a wire dipole antenna in detail, we note that, see Figure 1.4, radiation does *not* take place in directions along the charged particle acceleration.

Next, we will take a look at the radiation from a short dipole antenna.

1.2.2 Short Wire Dipole Radiation

We consider two short – that is much shorter than a wavelength – electrically conducting wires, each folded back 90 degrees, and connected to an AC source. We will look at the electric field around this structure at different instances of time within one half of the period T, see Figure 1.5.

- $t = 0^+$, **see Figure 1.5(a)**. The time is a short while after t = 0. The source has been turned on and charge is accelerated from the source to the wire ends. Because of the accelerating charges at the feed point, a transverse electric field component is traveling outward, in a direction perpendicular to the wires. Since field lines have to be continuous and start and end perpendicular to a charged body, the electric field line takes the form as shown. Underneath the dipole, the current is shown as a function of time; the time of the snapshot (0^+) is indicated with a black dot.
- $t = \left(\frac{T}{4}\right)$, **see Figure 1.5(b)**. At this moment, the current has reached its maximum value, its change with time has become zero. The electric field lines are as shown in the figure. The transverse electric field component that was created at $t = 0^+$ has traveled a distance of a quarter of a wavelength. New transverse electric field components have been created after the creation of this first one.
- $t = \left(\frac{T}{4}\right)^+$, **see Figure 1.5(c)**. The current has become less than the maximum value and the time change of the current has changed sign. Charges are now accelerated into

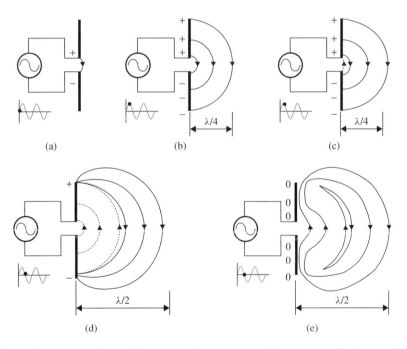

Figure 1.5 Electromagnetic radiation by charges in oscillatory acceleration. (a) $t = 0^+$. (b) $t = \left(\frac{T}{4}\right)$. (c) $t = \left(\frac{T}{4}\right)^+$. (d) $t = \left(\frac{T}{2}\right)^-$. (e) $t = \left(\frac{T}{2}\right)$.

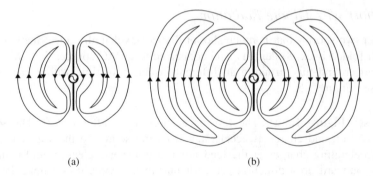

(a)   (b)

Figure 1.6 Detachment of electric field lines from a dipole antenna at different times. (a) $t = \left(\frac{T}{2}\right)^{+}$. (b) $t = T^{+}$.

the opposite direction and new electric field lines, oppositely directed relative to the existing ones, may be thought of as being created.

- $t = \left(\frac{T}{2}\right)^{-}$, **see Figure 1.5(d)**. The current amplitude has become very small and excess charges are only present at the dipole tips. Additional, upward-directed, transverse field lines have been created since $t = \left(\frac{T}{4}\right)^{+}$. The first one of these has traveled a distance of nearly a quarter of a wavelength.
- $t = \left(\frac{T}{2}\right)$, **see Figure 1.5(e)**. Both halves of the dipole antenna have become charge free. No excess charge is present and the current has become zero. The electric field lines do not need to be perpendicular to the conductors anymore, since these conductors have become charge-free. As a consequence, the field lines form closed loops and detach from the conductors.

For clarity reasons we have shown the mechanism of radiation from a dipole in a plane and only at one side of the dipole. Of course, in this plane, the radiation takes place at both sides, see Figure 1.6. In three dimensions, the field line pattern is rotationally symmetric around the axis of the dipole antenna. For the same clarity reasons we have left out magnetic field lines in our explanation. The magnetic field lines form closed loops around and perpendicular to the electric field lines.

Most wire antennas may be thought to consist of an infinite number of elementary (that is: infinitely small) dipole antennas.

With electromagnetic radiation and dipole antenna radiation now explained, we can continue with our short overview of antenna history.

1.3 The Modern History of Antennas

Guglielmo Marconi grasped the potential of Hertz's equipment and started experimenting with wireless telegraphy. In 1895 he hit upon a new arrangement of his equipment that suddenly allowed him to transmit and receive over distances that progressively increased up to and beyond 1.5 km [2, 3]. Marconi had enlarged the antenna. His monopole antenna,[4]

---

[4] A monopole forms a dipole antenna together with its image in the ground.

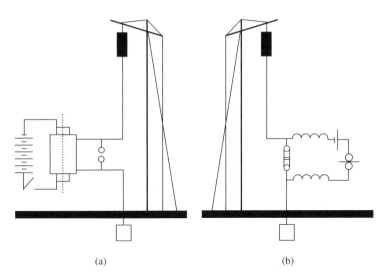

(a)   (b)

Figure 1.7 Marconi's antennas in 1895. (a) Scheme of the transmitter used by Marconi at Villa Griffone. (b) Scheme of the receiver used by Marconi at Villa Griffone. From [1].

see Figure 1.7, was resonant at a wavelength much larger than any that had been studied before and it was this creation of long-wavelength electromagnetic waves that turned out to be the key to his success. It was also Marconi who, in 1909, introduced the term *antenna* for the device that was formerly referred to as an *areal* or an *elevated wire* [3, 4].

The invention of the *thermionic valve* or *diode* by Fleming in 1905 and the *audion* or *triode* by Lee de Forest in 1907 paved the way for a reliable detection, reception and amplification of radio signals. From 1910 onwards broadcasting experiments were conducted that resulted in Europe in 1922 in the forming of the British Broadcasting Corporation (BBC) [5].

In the 1930s a return of interest to the higher end of the radio spectrum took place. This interest intensified with the outbreak of World War II with the immediate need for compact communication equipment as well as compact (airborne), high-resolution radar. Antenna design became a new specialism. At the end of World War II, antenna theory was mature to a level that made the analysis possible of, among others, free-standing dipole, horn and reflector antennas, monopole antennas, slots in waveguides and arrays thereof. The end of the War also saw the beginning of the development of electronic computers. The introduction of the IBM-PC[5] in 1981 considerably helped in the development of numerical electromagnetic analysis software. The 1980s may be seen as the decade of numerical microwave circuit and planar antenna theory development. In this period the numerical electromagnetics code (NEC), for the analysis of wire antennas, was commercially distributed. The 1990s, however, may be seen as the decade of the numerical electromagnetic-based microwave circuit and (planar, integrated) antenna design. In 1989 Sonnet started distribution, followed in 1990 by HP

---

[5] 4.77 MHz, 16 kb RAM, no hard drive.

(now Agilent), High Frequency Structure Simulator (HFSS).[6] [6] These two numerical electromagnetic analysis tools were followed by Zeland's IE3D, Remcom's XFdtd, Agilent's Momentum, CST's Microwave Studio, FEKO from EM Software & Systems, and others.

Despite the diversity of the numerical electromagnetic analysis software commercially available today, it is the author's strong belief that there exists, and will continue to exist, a need to develop an understanding of electromagnetics and antenna theory. At least a basic understanding needs to be present to be able to evaluate the results obtained with numerical electromagnetic analysis software. And a bit more than a basic understanding is necessary for designing new antennas. In the design process, the choice for a particular type of antenna is mainly dictated by the volume available for the antenna, the frequency (directly related to this volume) and the distance over which wireless communication needs to be performed.

1.4   Frequency Spectrum and Antenna Types

The radio frequency (RF) radiation of electromagnetic waves is used for frequencies that lie roughly between 30 Hz and 300 GHz. Table 1.1 [7] lists a number of frequency bands, associated wavelengths[7] and applications for these bands.

In the table we have made use of the IEEE-defined frequency band designations [8]. We stop at a frequency of 300 GHz, where infrared starts, followed by visible light from 400 THz upwards.

Of the many antenna types that exist, only a few basic ones will be discussed in detail in this book. Figure 1.8 shows these antenna types.

Other antenna types may be seen as combinations of these basic antennas (e.g. Yagy-Uda antennas are combinations of active and short-circuited dipole antennas placed in parallel) or derivatives of one of the basic antennas (e.g. a microstrip patch antenna may be seen as consisting of two rectangular aperture antennas). The detailed discussion of most of these antennas is beyond the scope of this short course in antenna theory and may be found in specialist textbooks, see for example [9].

1.4.1   Dipole Antennas

The short dipole antenna consists of two wires or circular tubes having a length much shorter than the wavelength and placed along the same axis, see Figures 1.5 and 1.6. The dipole antenna is voltage-excited in the small gap between the two dipole halves. This may be accomplished through a transmission line, connecting the gap to the voltage source. Short dipoles are used for radio broadcasting systems at VHF frequencies and below.

The resonant dipole antenna is an antenna where the length of the two wires or tubes together is a multiple (most often *one*) of half the wavelength. Half-wavelength dipole antennas are used for small-band applications at low GHz frequencies.

---

[6] Currently Ansoft HFSS.
[7] $\lambda = c/f$, where λ is the wavelength, c is the velocity of light and f is the frequency.

Table 1.1   RF Frequency band designations, wavelengths and applications

Frequency band	Frequencies	Wavelengths	Applications
extremely low frequency (ELF)	30–300 Hz	1000–10,000 km	Submarine communications
infralow frequency (ILF)	0.3–3 kHz	100–1000 km	–
very low frequency (VLF)	3–30 kHz	10–100 km	Navigation Weather
low frequency (LF)	30–300 kHz	1–10 km	Navigation Maritime communications Information and weather systems Time systems
medium frequency (MF)	0.3–3 MHz	0.1–1 km	Navigation AM radio Mobile radio
high frequency (HF)	3–30 MHz	10–100m	Citizen's band (CB) Shortwave radio Mobile radio Maritime radio
very high frequency (VHF)	30–300 MHz	1–10m	Amateur radio VHF TV FM radio Mobile satellite Mobile radio Fixed radio
ultra high frequency (UHF)	0.3–3 GHz	0.1–1m	Microwave Satellite UHF TV Paging Cordless telephony Cellular telephony Wireless LAN
super high frequency (SHF)	3–30 GHz	1–10 cm	Microwave Satellite Wireless LAN
extremely high frequency (EHF)	30–300 GHz	1–10 mm	Microwave Satellite Radiolocation

1.4.2   Loop Antennas

Small loop antennas may be considered to be *magnetic* dipole antennas. The fields from a small loop antenna are similar to those from a small dipole antenna with the electric and magnetic fields interchanged, as we will see in a later chapter. Loop antennas are used, among others, in direction finding systems.

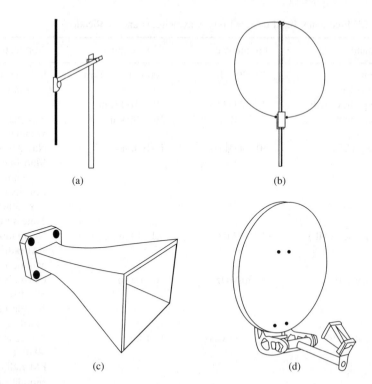

(a)   (b)

(c)   (d)

Figure 1.8 Basic antenna types. (a) Dipole antenna. (b) Loop antenna. (c) Horn antenna. (d) Parabolic reflector antenna.

1.4.3   Aperture Antennas

An aperture antenna consists of an 'opening' in a metallic surrounding. The fields across this opening, that is the aperture, radiate into free space. An electromagnetic horn is an antenna where the radiating aperture is matched to the waveguiding system that supports the excitation signal. This matching is accomplished through properly shaping the transition from waveguiding structure to aperture. Aperture antennas are used in the GHz range of frequencies.

1.4.4   Reflector Antennas

A reflector antenna is also an aperture antenna. A primary radiator (dipole or horn antenna) in the focal point of a parabolic reflector illuminates the reflector. The aperture formed by this reflector then radiates into free space. Since the radiated waves are concentrated into a beam, the width of which is inversely proportional to the size of the aperture as we will see in a later chapter, reflector antennas offer a convenient way to concentrate radiation. This allows long distance communication in the low and high GHz range of frequencies. Parabolic reflector antennas are used, among others, for satellite television transmission and reception, and for radar (radio detection and ranging).

1.4.5 Array Antennas

Antennas may be combined with similar or other types of antenna to form an array antenna. The antennas within the array we designate as *antenna elements* or *array elements* or, for short, *elements*. The combination of all the elements we designate *array antenna*, *antenna array* or *array*. Commonly, similar elements are positioned at regular intervals on a line (linear array antenna) or in a plane (planar array antenna). By forming an array, a radiation beam may be created having a small beamwidth. By electronically controlling the phase differences between the elements, we may electronically direct the beam in different directions without physically rotating the antenna. This possibility of electronic beam steering or *scanning* makes (phased) array antennas particularly attractive.

1.4.6 Modern Antennas

In this book we will briefly touch upon the subject of modern antennas. By 'modern' we mean antennas that can be considered as derivatives of the basic antennas that are dealt with in detail. Of all modern antennas we will, briefly, discuss the printed monopole antenna, the inverted F antenna (IFA) and the microstrip patch antenna, see Figure 1.9. These antennas may be encountered in today's wireless devices.

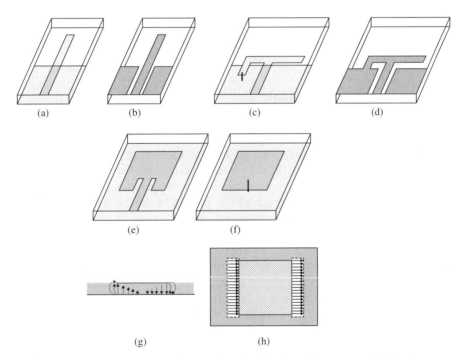

Figure 1.9 Modern antennas. (a) Microstrip printed monopole antenna. (b) Coplanar waveguide (CPW) printed monopole antenna. (c) Microstrip IFA. (d) CPW IFA. (e) Microstrip excited microstrip patch antenna. (f) Probe excited microstrip patch antenna. (g) Side view of electric field lines patch antenna. (h) Top view of electric field lines patch antenna.

The printed monopole antenna, either in microstrip technology (with a separate trace and a ground layer, see Figure 1.9(a) or in CPW technology (with a combined trace and ground layer, see Figure 1.9(b)), may be regarded as an asymmetric printed dipole antenna, the arms being the monopole and the ground. The asymmetry causes differences in the current distributions in the two arms which results in some disturbances in the dipole radiation pattern. To avoid the existence of multiple lobes, the length of the ground plane should remain smaller than a quarter of a wavelength.

The inverted F antenna (IFA) may be realized in microstrip technology, see Figure 1.9(c), or in CPW technology, see Figure 1.9(d). The IFA may be regarded as a printed monopole antenna where the top section has been folded down. The folded-down part, being parallel to the ground plane, introduces capacitance to the input impedance. This additional capacitance is compensated for by a short circuited stub. The realization of the short circuit is straight forward in CPW technology, but requires a via to the ground plane in microstrip technology, see Figure 1.9(c).

A microstrip patch antenna, see Figures 1.9(e), 1.9(f), may be considered as a cavity with electrical conducting top and bottom and magnetically conducting side walls. The fields inside the 'cavity' may be excited by, among others, a microstrip transmission line, see Figure 1.9(e), or a probe, see Figure 1.9(f). Since the walls of this 'cavity' are not perfectly conducting, the electric fields will 'fringe' at the edges, see Figure 1.9(g), and the horizontal components of these so-called 'fringe fields' are responsible for the radiation. If the length of the patch is chosen to be half a wavelength in the dielectric sheet, the radiation may be thought to originate from two in-phase slots as depicted in Figure 1.9(h).

1.5 Organization of the Book

Before we dive into the field calculations of different antenna types, we will first introduce, in Chapter 2, the so-called *performance parameters* of antennas. These parameters serve to evaluate antennas, offer a means to compare them to each other and offer the possibility of including antenna performance in a high-level system evaluation. Since a thorough understanding of these parameters is considered to be of paramount importance for the practical antenna engineer, we designate a large portion of this book to this topic.

With the antenna parameters well understood, the derivation of these parameters may be undertaken for a selected group of antennas. Before we do so, however, we think it is wise to first give a short refresher course in vector algebra. This is the topic of Chapter 3. Then, in Chapter 4, we introduce the calculation of far-fields from general current distributions and introduce the concept of reciprocity. In Chapter 5, we will use the concepts developed in Chapter 4 to calculate the fields of an elementary dipole antenna and a half-wave dipole antenna. In Chapter 6 we will analyze the loop antenna. Chapter 7 is devoted to aperture antennas, and the theory developed there will be applied to a horn antenna, a parabolic reflector antenna and a rectangular microstrip patch antenna. Finally, in Chapter 8 we will introduce array antennas.

Most chapters are concluded with a more practically oriented section wherein the acquired knowledge is used for designing a modern (mobile) antenna using CST Microwave Studio® (CST MWS), a commercially available specialist tool for the

3D EM simulation of high frequency components [10]. The steps, outcomes and resulting next iterations are explained so that having access to this software suite is not absolutely necessary.

1.6 Problems

1.1 What is an antenna?

1.2 What is the source of electromagnetic radiation?

1.3 Why do point source electromagnetic radiators not exist in reality?

References

1. H.J. Visser, *Array and Phased Array Antenna Basics*, John Wiley & Sons, Chichester, UK, 2005.
2. G. Masini, *Marconi*, Marsilio Publishers, New York, 1995.
3. G. Marconi, 'Wireless Telegraphic Communications', *Nobel Lectures in Physics*, 1901–21, Elsevier, 1967.
4. G. Pelosi, S. Selleri and B. Valotti, 'Antennae', *IEEE Antennas and Propagation Magazine*, Vol. 42, No. 1, February 2000, pp. 61–63.
5. J. Hamilton (ed.), *They Made Our World, Five Centuries of Great Scientists and Inventors*, Broadside Books, London, 1990, pp. 125–132.
6. J.C. Rautio, 'Planar Electromagnetic Analysis', *IEEE Microwave Magazine*, March 2003, pp. 35–41.
7. R. Horak, *Webster's New World Telecom Dictionary*, John Wiley & Sons, New York, 2007.
8. IEEE Standards Board, 'IEEE Standard Letter Designations for Radar-Frequency Bands', *IEEE Standard IEEE Std 521-1984*, March 1984.
9. C.A. Balanis, *Antenna Theory, Analysis and Design*, *third edition*, John Wiley & Sons, New York, 2005.
10. CST Computer Simulation Technology: http://www.cst.com

2

Antenna System-Level Performance Parameters

Now that we have developed a basic understanding of the operation of antennas, based on physical reasoning, it is time to quantify and interrelate the different antenna parameters that describe antenna functioning. Therefore, in this chapter we will introduce these system-level performance parameters for antennas.[1] With these parameters we will be able to calculate the effect of an antenna or antennas in a communication or radar system.

2.1 Radiation Pattern

In the previous chapter we saw that accelerating charge – and thus displacement of charge – is the cause of electromagnetic radiation. Due to this displacement, every antenna must have a non-zero size. We have already seen that even the smallest-size antenna produces an electromagnetic radiation (propagating field disturbance) that is not uniform (i.e. not equally distributed in all directions), see Figures 1.4–1.6. This non-uniform radiation is described by the so-called *radiation pattern*, which is evaluated in the *far-field*.

Although uniformly radiating antennas or *isotropic radiators* cannot exist in real life, they may come in handy for comparing different antennas. The uniform radiator must then be seen as a mathematical abstraction.

---

[1] Portions of text in this Chapter have been reproduced from: Visser, H. '*Array and Phased Array Antenna Basics*'. This includes the following text: For the half wave (p. 17)...with respect to the main lobe (p. 18); In figure 2.5 (p. 20)...a matter of taste (p. 22); So far we have (p. 25)...into other directions (p. 26); The directivity (p. 26)...losses in the antenna (p. 27); In comparing (p. 28)...need larger antennas (p. 46). Reproduced with permission from John Wiley & Sons, Ltd. This is presented in [1].

Antenna Theory and Applications, First Edition. Hubregt J. Visser.
© 2012 John Wiley & Sons, Ltd. Published 2012 by John Wiley & Sons, Ltd.

2.1.1   Field Regions

When talking about radiated fields, we have to take into account the distance relative to the antenna where these fields are evaluated. Close to the antenna, a region exists where energy is stored and returned to the antenna. This region is called the *reactive near-field region* of the antenna [2, 3]. Moving away from the antenna, through the reactive near-field region, the next region encountered is called the *radiating near-field region* or *Fresnel region* [2, 3]. The radiating near-field region is characterized by the fact that the radiation fields dominate the reactive fields and that the angular distribution of this radiated field is dependent on the distance from the antenna.

The *far-field region* is that region where not only the radiating fields predominate, but also the angular field distribution has become independent of the distance from the antenna [2, 3]. In the far-field region, the electric field vector **E** and the magnetic field vector **H** are perpendicular to the observation direction **r** and to each other, see Figure 2.1. Power flow in the far-field region is therefore only in the direction of **r**, in contrast to the situation in both near-field regions.

The radiated fields of an antenna will be evaluated in the far-field region. In real-life situations, the separation between transmitting antenna and receiving antenna will (almost) always be such that the antennas are in each other's far-field regions. As a rule of thumb, the far-field is the region for which

$$r \geq \frac{2D^2}{\lambda},$$

(2.1)

where D is the largest dimension of the antenna and λ is the wavelength used.

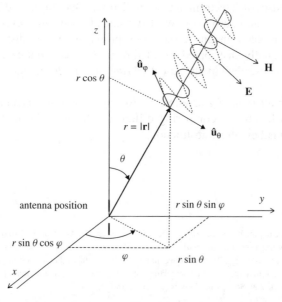

Figure 2.1   An antenna in a rectangular coordinate system, the relation between rectangular and spherical coordinates and the radiated electric and magnetic far-fields. $\hat{\mathbf{u}}_\vartheta$ and $\hat{\mathbf{u}}_\varphi$ are unit vectors in the ϑ- and φ-direction, respectively.

As we will see in a later chapter, the equality sign in equation (2.1) corresponds to a maximum phase error of 22.5°.

Although we have thus indicated a sharp boundary between the near- and far-field regions, in practice these boundaries are not that sharp or precisely determined.

We will now look at the phase differences stemming from different parts of an antenna that will cause constructive interference in some directions and destructive interference in other directions, knowing that we have to evaluate these effects in the far-field region.

2.1.2 Three-Dimensional Radiation Pattern

For the half-wave dipole antenna we have seen, in the previous chapter, that no radiation occurs in directions along the dipole axis. To explain this in more detail, we consider a horizontal half-wave dipole that we represent by two (hypothetic) isotropic radiators, 1 and 2 in Figure 2.2, spaced apart by half a wavelength. We will have a look at the electric field in the far-field region, for the three situations, depicted I, II and III in the Figure.[2]

- **Situation I.** In the direction along the dipole axis, the waves emitted by isotropic radiators 1 and 2 are 180°. out of phase and therefore cancel each other out. There is no radiation in directions along the dipole axis.
- **Situation II.** In the direction perpendicular to the dipole axis, the waves emitted by isotropic radiators 1 and 2 are in phase and add, giving the maximum possible amplitude. When we evaluate these signals at infinity – what we should do, strictly speaking, to have a far-field region evaluation – the two distinct directions stemming from isotropic radiators 1 and 2 become one.
- **Situation III.** For a direction in between parallel and perpendicular to the dipole axis, the two waves add, but with a phase difference due to the fact that one wave reaches a far-field evaluation point before the other one does. Therefore, the amplitude of the combined waves will be less than maximum.

In Figure 2.2, we have restricted ourselves to the evaluation in the plane parallel to the dipole axis and containing the dipole. If we evaluate the electric far-field amplitude for all possible angular positions ϑ, φ, the three-dimensional *radiation pattern* of Figure 2.3 results. Note that this pattern is for a half-wave dipole antenna directed along the Cartesian z-axis, also indicated in the figure. Contour plots of the radiation pattern (equi-amplitude lines) are shown, projected on the x,y-plane.

We see that the radiation pattern is symmetrical around the dipole axis, at maximum in the directions perpendicular to the dipole axis and zero in the directions along the dipole axis.

Although we are dealing here with a basic antenna having a relatively simple radiation pattern, we already encounter some difficulty in reading and interpreting this radiation pattern. This will become worse for more complicated antennas. To illustrate this we have created the radiation pattern shown in Figure 2.4, which could be the pattern of a pyramidal horn antenna. The position and orientation of this horn is indicated in the figure.

---

[2] We could also have chosen to look at the magnetic field since the magnetic field is perpendicular to, and in phase with, the electric field.

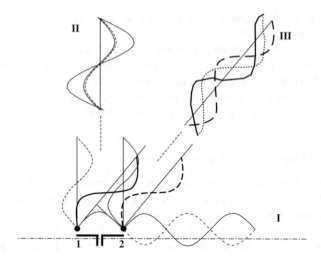

Figure 2.2   A half-wave dipole is represented by two isotropic radiators, 1 and 2, spaced apart by half a wavelength.

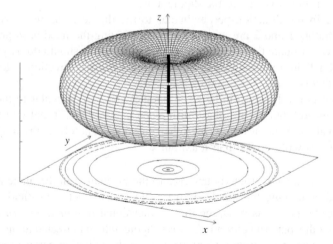

Figure 2.3   Three-dimensional electric field radiation pattern of a z-directed half-wave dipole antenna.

Although we encounter even more difficulties in the interpretation of this pattern, we can still distinguish some salient features of this pattern. First of all we observe angular regions of strong radiation, surrounded by regions of weak radiation. We call these regions of strong radiation *lobes*, see 1, 2 and 3 in Figure 2.4. The biggest lobe, number 1 in Figure 2.4, is called the *main lobe* or *main beam*. The main lobe contains the direction of maximum radiation. Here, that direction is given (in spherical coordinates) by $\vartheta = 0$. All lobes other than the main lobe are called *minor lobes* or *side lobes*. Side lobes, numbers

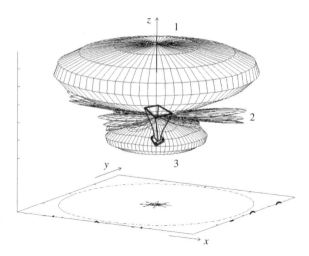

Figure 2.4 Three-dimensional electric field radiation pattern of a hypothetical pyramidal horn antenna.

2 and 3 in Figure 2.4, are radiation lobes pointing in directions other than the main lobe. The side lobe number 3 in the figure is also called a *back lobe*, due to its position relative to the antenna, that is 180° rotated with respect to the main lobe.

It is clear that the three-dimensional plot serves best for inspection purposes, for example the identification of side lobes and their angular distribution. It is not easy to obtain qualitative information from this representation. For that last purpose it is customary to use planar cuts of the three-dimensional radiation patterns.

2.1.3 Planar Cuts

In Figure 2.5 a planar cut of the three-dimensional radiation pattern of a half-wave dipole antenna, see Figure 2.3, is shown.

This normalized, planar cut, transformed to the two-dimensional domain, is shown in Figure 2.6.

The elevation angle ϑ increases, going clockwise around the circle. The amplitude of the electric field is plotted along the radius of the circle.

Normally, the azimuth angle φ where the cut is taken should be specified, but since in this particular case we are dealing with a radiation pattern that is rotationally symmetric, all cuts are identical.

A planar cut of the (artificial) radiation pattern of the hypothetical pyramidal horn antenna, see Figure 2.4, is shown in Figure 2.7.

This normalized, planar cut, transformed to the two-dimensional domain, is shown in Figure 2.8.

Although, all information can be read from the polar plots and the polar plots correspond to our perception of the physical three-dimensional world, a rectangular plot may be found helpful in observing the details in the side lobes.

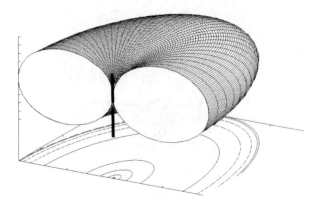

Figure 2.5   Planar cut from the three-dimensional radiation pattern of a half-wave dipole antenna shown in Figure 2.3.

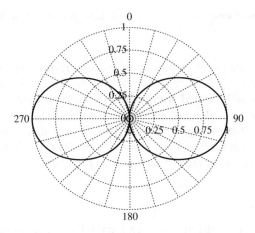

Figure 2.6   Normalized polar plot of the electric field radiation pattern of a half-wave dipole antenna.

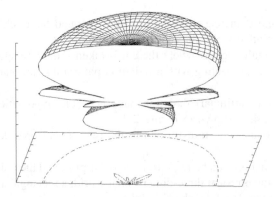

Figure 2.7   Planar cut from the three-dimensional radiation pattern of a hypothetical pyramidal horn antenna as shown in Figure 2.4.

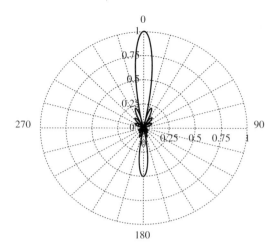

Figure 2.8 Normalized, polar plot of the electric field radiation pattern of a hypothetical pyramidal horn antenna.

To construct rectangular plots, we return to the three-dimensional domain, where we now represent the field amplitude in a different format. In the horizontal plane, we now represent ϑ and φ in a polar format: ϑ varies between $0°$ and $180°$ going from the center of the circle to the outer rim. φ varies between $0°$ and $360°$, following the circle contour. The amplitude of the radiated electric far-field is plotted along the vertical axis. Figures 2.9 and 2.10 represent the radiation patterns according to this format.

By taking planar cuts through planes $\varphi = constant$, we obtain the desired rectangular plots – with the angle ϑ along the horizontal axis and the electric field amplitude along the vertical axis – as shown in Figures 2.11 and 2.12.

The correspondence between rectangular and polar plots is shown in Figure 2.13.

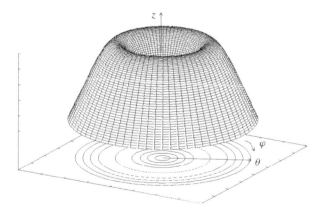

Figure 2.9 Alternative three-dimensional electric field radiation pattern of a half-wave dipole antenna.

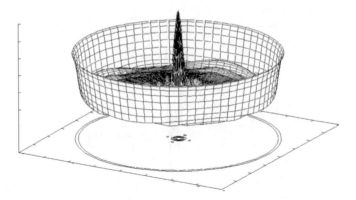

Figure 2.10 Alternative three-dimensional electric field radiation pattern of a hypothetical pyramidal horn antenna.

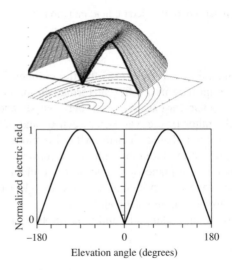

Figure 2.11 Planar cut from the three-dimensional radiation pattern of a half-wave dipole antenna as shown in Figure 2.9.

Whether a polar or a rectangular plot of the radiation pattern is used, is mainly a matter of taste. The author's taste is for rectangular patterns, especially for observing details, and for *power patterns* over *field patterns*.

2.1.4   Power Patterns

The radiation patterns shown thus far have been so-called *field patterns*. In general, we are not directly interested in the angular electric field amplitude distribution, but more in the angular power distribution. In the far-field region, the power is related to the electric field through a square-law relationship.

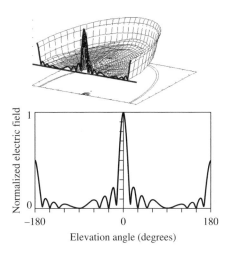

Figure 2.12 Planar cut from the three-dimensional radiation pattern of a hypothetical pyramidal horn antenna as shown in Figure 2.10.

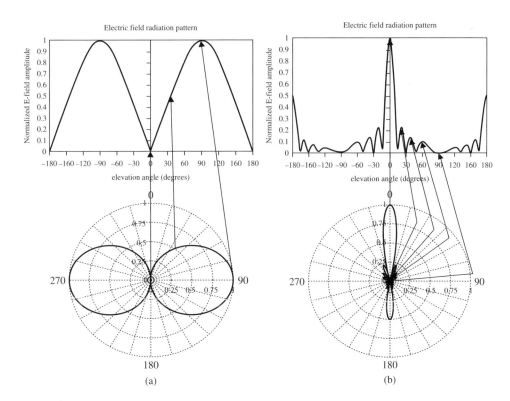

Figure 2.13 Correspondence between polar and rectangular radiation patterns. (a) Half-wave dipole antenna. (b) Pyramidal horn antenna.

To demonstrate this, we start by stating that the electric far-field of an antenna, in polar coordinates, is given by

$$E_\vartheta(\mathbf{r}) = E_\vartheta(\vartheta, \varphi) \frac{e^{jk_0 r}}{r}, \tag{2.2}$$

$$E_\varphi(\mathbf{r}) = E_\varphi(\vartheta, \varphi) \frac{e^{jk_0 r}}{r}, \tag{2.3}$$

$$E_r(\mathbf{r}) = 0, \tag{2.4}$$

where r is the distance, ϑ and φ are elevation and azimuth angle, respectively, and $k_0 = 2\pi/\lambda$ is the wave number. In a later chapter these equations will be derived, as well the relation between magnetic and electric field in the far-field of an antenna:

$$\mathbf{H}(\mathbf{r}) = \frac{1}{Z_0} \hat{\mathbf{u}}_r \times \mathbf{E}(\mathbf{r}), \tag{2.5}$$

where $Z_0 = \sqrt{\mu_0/\varepsilon_0}$ is the characteristic impedance of free space.

The power density or Poynting's vector $\mathbf{S}(\mathbf{r}, t)$, radiated by the antenna at time t in the direction $\mathbf{r}$, is given by

$$\mathbf{S}(\mathbf{r}, t) = \mathbf{E}(\mathbf{r}, t) \times \mathbf{H}(\mathbf{r}, t) = \text{Re}\left[\mathbf{E}(\mathbf{r})e^{j\omega t}\right] \times \text{Re}\left[\mathbf{H}(\mathbf{r})e^{j\omega t}\right]$$
$$= \frac{1}{2}\left[\mathbf{E}(\mathbf{r})e^{j\omega t} + \mathbf{E}^*(\mathbf{r})e^{-j\omega t}\right] \times \frac{1}{2}\left[\mathbf{H}(\mathbf{r})e^{j\omega t} + \mathbf{H}^*(\mathbf{r})e^{-j\omega t}\right]$$
$$= \frac{1}{4}\left[\mathbf{E}(\mathbf{r}) \times \mathbf{H}(\mathbf{r})e^{j2\omega t} + \mathbf{E}^*(\mathbf{r}) \times \mathbf{H}^*(\mathbf{r})e^{-j2\omega t} + \mathbf{E}(\mathbf{r}) \times \mathbf{H}^*(\mathbf{r}) + \mathbf{E}^*(\mathbf{r}) \times \mathbf{H}(\mathbf{r})\right]. \tag{2.6}$$

The time-averaged Poynting vector $\mathbf{S}(\mathbf{r})$ over one period $T = 2\pi/\omega$ is then found to be

$$\mathbf{S}(\mathbf{r}) = \frac{1}{T}\int_0^T \mathbf{S}(\mathbf{r}, t)dt = \frac{1}{4}\left[\mathbf{E}(\mathbf{r}) \times \mathbf{H}^*(\mathbf{r}) + \mathbf{E}^*(\mathbf{r}) \times \mathbf{H}(\mathbf{r})\right] = \frac{1}{2}\text{Re}\left[\mathbf{E}(\mathbf{r}) \times \mathbf{H}^*(\mathbf{r})\right]. \tag{2.7}$$

Substitution of equation (2.5) in equation (2.7) leads to

$$\mathbf{S}(\mathbf{r}) = \frac{1}{2}\text{Re}\left[\mathbf{E}(\mathbf{r}) \times \mathbf{H}^*(\mathbf{r})\right] = \frac{1}{2Z_0}\text{Re}\left[\mathbf{E}(\mathbf{r}) \times \left(\hat{\mathbf{u}}_r \times \mathbf{E}^*(\mathbf{r})\right)\right]$$
$$= \frac{1}{2Z_0}\text{Re}\left[\mathbf{E}(\mathbf{r}) \cdot \mathbf{E}^*(\mathbf{r})\hat{\mathbf{u}}_r - \mathbf{E}(\mathbf{r}) \cdot \hat{\mathbf{u}}_r \mathbf{E}^*(\mathbf{r})\right]$$
$$= \frac{1}{2Z_0}|\mathbf{E}(\mathbf{r})|^2 \hat{\mathbf{u}}_r. \tag{2.8}$$

In the above equation use has been made of $\mathbf{E}(\mathbf{r}) \cdot \hat{\mathbf{u}}_r = 0$. This follows from equations (2.2)–(2.4), that state that the electric field is perpendicular to $\hat{\mathbf{u}}_r$. The vector algebra used in equation (2.8) will be reviewed in the next chapter.

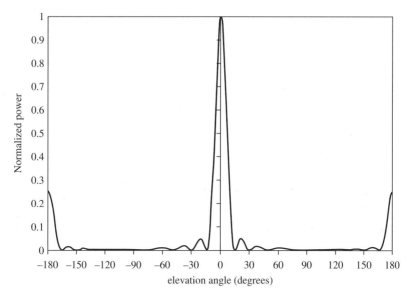

Figure 2.14   Power radiation pattern for a hypothetical pyramidal horn antenna.

So by taking the square of the electric field amplitude of the normalized field pattern, we may obtain the normalized *power pattern*. The normalized power pattern (in the plane $\varphi = 0$) for the hypothetical pyramidal horn antenna is shown in Figure 2.14.

We see that by plotting the power instead of the field amplitude, we seem to have lost the detailed information in the side lobe region. Ideally, we would want to observe the same detail of information in the side lobes as in the main lobe. This may be accomplished by plotting the angular power distribution not on a linear scale (as we have done in Figure 2.14) but on a *logarithmic* scale. In Figure 2.15 the power pattern is shown in *decibels*, where $x(\mathrm{dB}) = 10\log(x)$.

2.1.5   Directivity and Gain

So far we have been comparing the radiated field and power only to the maximally radiated field or power of the same antenna. If we want to compare different antennas with each other, we need to have a reference to compare them to. This reference is taken to be the isotropic radiator.

Although we know that the isotropic radiator is a physical abstraction, it still may serve as a reference for real life antennas. The (hypothetical) isotropic radiator, radiates equally in all directions. Its normalized, three-dimensional electric field or power pattern, on a linear scale, therefore is a sphere with radius *one*.

The *directivity function*, $D(\vartheta, \varphi)$, is defined as the power radiated by an antenna in a direction (ϑ, φ), compared to the power radiated in that same direction by an isotropic radiator.

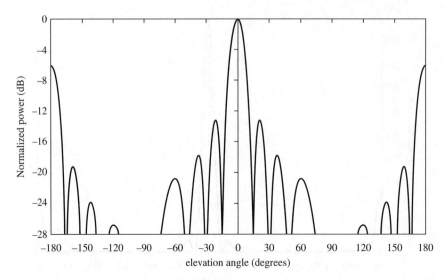

Figure 2.15 Power radiation pattern for a hypothetical pyramidal horn antenna on a logarithmic (decibel) scale.

The power radiated in a certain direction by an isotropic radiator is equal to the amount radiated into every other direction and is equal to $P_t/4\pi$, where P_t is the total transmitted power and $P_t/4\pi$ therefore is the radiation intensity of the isotropic radiator.[3] Thus

$$D(\vartheta, \varphi) = \frac{P(\vartheta, \varphi)}{P_t/4\pi}, \tag{2.9}$$

where $P(\vartheta, \varphi)$ is the power radiated by the actual antenna in the direction (ϑ, φ).

So, we compare the power radiated by the actual antenna to the power that would have been radiated by an isotropic radiator, radiating the same total amount of power.

In Figure 2.16 we show both the three-dimensionally radiated power by a half-wave dipole and the radiated power by an isotropic radiator. Both antennas – the real half-wave dipole antenna and the hypothetical isotropic radiator – have the same total transmitted power.

The figure clearly shows the value of using the isotropic radiator as a reference. We see that in some directions, the half-wave dipole antenna radiates more power than the isotropic radiator while for other directions the opposite is true. Since both antennas have the same amount of total transmitted power, we may transform the three-dimensional radiation pattern of one antenna into that of the other by reducing power in certain directions and increasing power into other directions.

The *directivity*, D, is defined as the maximum of the directivity function:

$$D = \max[D(\vartheta, \varphi)]. \tag{2.10}$$

---

[3] 4π is the value of the solid angle of a complete sphere. Compare this three-dimensional angle with an angle in the two-dimensional domain, where 2π is the angle of a complete circle.

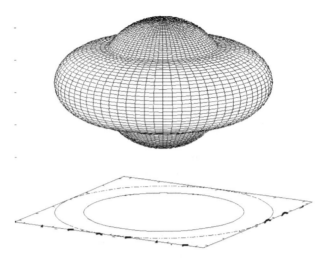

Figure 2.16 Three-dimensional, normalized power radiation pattern of a (hypothetical) isotropic radiator and a half-wave dipole antenna.

For the half-wave dipole antenna, the directivity is $D = 1.64$ as we will see in a later chapter or $D = 2.15 \, \text{dBi}$, meaning that the half-wave dipole antenna at maximum radiates 1.64 times as much power as an isotropic radiator would do, transmitting the same total amount of power.

In general though, the total transmitted power of an antenna is not known, or is difficult to assess. Therefore, a second function exists: the *gain function*, $G(\vartheta, \varphi)$. The gain function resembles the directivity function, except for the total radiated power having been replaced by the total accepted power, P_{in},

$$G(\vartheta, \varphi) = \frac{P(\vartheta, \varphi)}{P_{in}/4\pi}. \tag{2.11}$$

P_{in} is easier to assess than the total radiated power.

The gain function does *not* take impedance mismatch on the antenna terminals into account. If 99% of the power delivered to the antenna terminals is reflected, the gain function tells us how the remaining 1% of this power (i.e. the accepted power) is distributed in space. The *gain*, G, is the maximum of the gain function,

$$G = \max[G(\vartheta, \varphi)]. \tag{2.12}$$

The quotient of gain and directivity equals the quotient of total radiated power and total accepted power and is called the *radiation efficiency*, η [3].

$$\eta = \frac{G}{D} = \frac{P_t}{P_{in}}. \tag{2.13}$$

The efficiency is smaller than one due to ohmic and/or dielectric losses in the antenna.

2.1.6 Antenna Beamwidth

In comparing antennas we may use the gain we just discussed. But this is a number that only tells us about the maximum radiation. Often we want to know the shape of the area of maximum radiation. For that purpose we use the *beamwidth*. The beamwidth tells us about the shape of the main lobe. Different beamwidth definitions exist. The two most often used beamwidths are shown in Figure 2.17.

The *half-power beamwidth*, ϑ_{HP}, is the angular separation between the points on a cut of the main lobe where the transmitted (received) power is half that of the maximum transmitted (received) power. This is shown in Figure 2.17(a). Since $10\log(0.5) = -3.01\,\text{dB} \approx -3\,\text{dB}$, ϑ_{HP} is found on a logarithmic scale on the interception points where the main lobe is $3\,\text{dB}$ under the maximum value, see Figure 2.17(b). The half-power beamwidth is therefore also known as the $3\,\text{dB}$ *beamwidth*.

Other less commonly used definitions for beamwidth are the *first null beamwidth*, ϑ_{FN}, indicated in Figure 2.17(a), 2.17(b) and the $10\,\text{dB}$ *beamwidth*. In general, when the term *beamwidth* is used, the $3\,\text{dB}$ *beamwidth* is meant.

Also indicated in Figure 2.17(b) is the level of the first and highest side lobe. This level is known as the *side-lobe level* (SLL).

The cuts are normally taken in the so-called *E-plane* and *H-plane*, also known as the *principal planes*. The *E*-plane is the plane that contains the electric field vector and the direction of maximum radiation. The *H*-plane is the plane that contains the magnetic field vector and the direction of maximum radiation. It is common practice to orient an antenna so that at least one of the principal plane patterns coincides with one of the geometrical

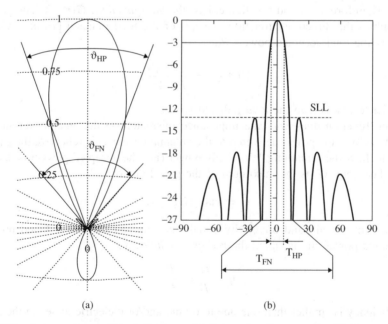

Figure 2.17 Antenna beamwidth definitions. (a) Polar plot, linear scale. (b) Rectangular plot, logarithmic scale.

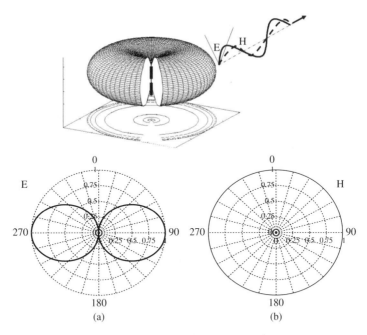

E

H

0

0.75

0.5

E

270   90

0.25 0.5 0.75

180

(a)

0

0.75

0.5

0.25

H

270   90

0.25 0.5 0.75

180

(b)

Figure 2.18   Normalized field patterns in the principal planes of a z-directed half-wave dipole antenna. (a) E-plane, containing the electric field vector and the direction of maximum radiation (figure-of-eight shape). (b) H-plane, containing the magnetic field vector and the direction of maximum radiation (circle with unit radius).

principal planes [2]. As an example, Figure 2.18 shows the normalized *field* patterns in the principal planes of a half-wave dipole antenna.

Note that for the z-directed half-wave dipole antenna, any plane $\varphi = constant$ is an E-plane.

2.2   Antenna Impedance and Bandwidth

Antennas will never be used as stand-alone devices; we will have to consider antennas as being part of a system. Therefore, besides the radiation characteristics, we also need to know the impedance characteristics.

When looking into the antenna, at the antenna terminals, we may regard the antenna as a complex impedance Z_A,

$$Z_A = R_A + jX_A, \tag{2.14}$$

where R_A is the real or resistive part of the antenna impedance and X_A is the imaginary or reactive part of the antenna impedance. The real part accounts for the dissipation and consists of two parts, the ohmic losses (which may be considerable in small antennas), R_L, and the (wanted) radiation losses, R_r,

$$R_A = R_L + R_r. \tag{2.15}$$

R_r is known as the *antenna radiation resistance*.

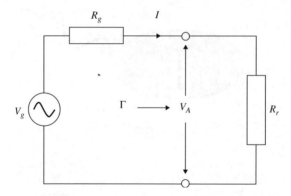

Figure 2.19   Equivalent circuit for matching the lossless antenna to the generator.

The reactive part of the antenna impedance accounts for the reactive near-field region of the antenna, where energy is being stored.

We assume for the moment that the reactive and ohmic loss part of the antenna impedance may be neglected ($Z_A = R_r$) and that the antenna is connected to a generator with (real) impedance R_g,[4] see Figure 2.19.

The (time average) power delivered by the generator to the antenna is given by

$$P = \frac{1}{2}\Re\{V_A \cdot I^*\}, \tag{2.16}$$

where $\Re\{x\}$ means the real part of the complex number x and I^* is the complex conjugate of I.

Since we are dealing with real quantities only, the power delivered to the antenna is

$$P = \frac{1}{2}|V_g|^2 \left|\frac{R_g}{R_g + R_r}\right|^2 \frac{1}{R_r}, \tag{2.17}$$

where use is made of

$$V_A = \frac{R_g}{R_g + R_r} V_g, \tag{2.18}$$

and

$$I^* = I = \frac{V_A}{R_r}. \tag{2.19}$$

To find the value of the antenna impedance that results in a maximum power transfer from the generator to the antenna, the following condition should be fulfilled:

$$\frac{\partial P}{\partial R_r} = 0. \tag{2.20}$$

---

[4] The principle of the so-called *conjugate matching* should be explained – in the most general form – for complex generator and antenna impedances, see for example [2] or [6]. For reasons of clarity, an explanation using real impedances is chosen here.

Upon substitution of equation (2.17) into equation (2.20), we finally find that the condition for maximum power transfer results in

$$R_r = R_g, \tag{2.21}$$

meaning that the radiation resistance should be equal to the generator resistance.

For the real generator and antenna impedance, this is also the condition for zero reflections at the antenna terminal. From (microwave) network theory, the reflection coefficient, Γ, looking into the antenna terminals, see Figure 2.19, is found to be

$$\Gamma = \frac{R_r - R_g}{R_r + R_g}. \tag{2.22}$$

When we design the antenna such that $R_r = R_g$, we see that the reflection becomes zero.

The impedance of an antenna normally varies as a function of the frequency, and therefore the matching also varies as a function of the frequency. This means that an antenna will only operate efficiently within a restricted band of frequencies. The width of this band of frequencies is called the *bandwidth*. Normally, at the center frequency (middle of the frequency band) the impedance matching will be best, and going to lower or higher frequencies results in a degradation of impedance matching up to a level where matching has become unacceptably poor. These levels determine the boundaries of the frequency band. Reflection coefficient levels of $-10\,\text{dB}$ ($|\Gamma|^2 \leq 0.1$) or $-15\,\text{dB}$ ($|\Gamma|^2 \leq 0.03$) are commonly employed to determine the *impedance bandwidth*. For relatively small bandwidth antennas, bandwidth is expressed in percentages of the center frequency [2]. If f_0 is the center frequency, f_l is the lower boundary of the frequency band and f_u is the upper boundary of the frequency band, the bandwidth is given by:

$$BW = \frac{f_u - f_l}{f_0} \cdot 100\%. \tag{2.23}$$

Bandwidths expressed in percentages of the center frequency are used up to a few decades. For larger bandwidth antennas the ratio of the upper and lower frequency boundaries is used (such as $10:1$ or $30:1$) [2].

As well as the *impedance bandwidth* a *radiation pattern bandwidth* may also exist. The two bandwidths need not be identical. It depends on the antenna and the application which of the two bandwidths is more critical.

Example

The reflection coefficient of a certain antenna is measured as a function of frequency and plotted (in dB) in Figure 2.20.

The reflection is at minimum at the center frequency, $f_0 = 11.3\,\text{GHz}$.

The $-10\,\text{dB}$-bandwidth then follows from the lower frequency limit $f_l = 10.97\,\text{GHz}$ and upper frequency limit $f_u = 11.93\,\text{GHz}$, both read from the graph, and equation (2.23)

$$BW_{-10\,\text{dB}} = \frac{11.93 - 10.97}{11.3} \cdot 100\% = 8.50\%. \tag{2.24}$$

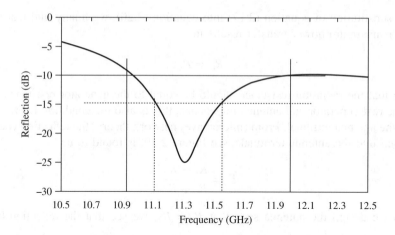

Figure 2.20   Measured reflection as a function of frequency.

The −15 dB-bandwidth follows from the graph and equation (2.23)

$$BW_{-15\,\text{dB}} = \frac{11.53 - 11.12}{11.3} \cdot 100\% = 3.63\%. \tag{2.25}$$

2.3   Polarization

The electric field in the far-field region of an antenna will in general possess two spherical coordinate components, E_ϑ and E_φ, see Figure 2.21.

In general, a phase difference will exist between these two field components. Therefore, the electric field *vector*, as a function of time, t, will – for an arbitrary phase difference – describe an ellipse in the ϑ,φ-plane. The electric field and the antenna are called *elliptically polarized*, see Figure 2.22(a).

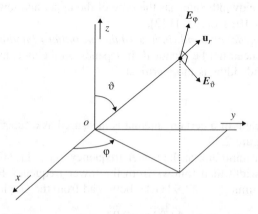

Figure 2.21   Electric field components E_ϑ and E_φ in the far-field region of an antenna placed at the origin O. $\mathbf{u}_r$ is the unit vector in the direction of wave propagation.

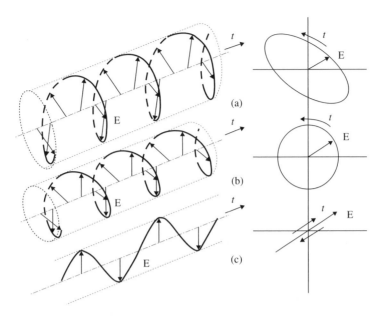

Figure 2.22 Polarization states. (a) Elliptical polarization. (b) Circular polarization. (c) Linear polarization.

When the phase difference is plus or minus 90° and the field components are equal in amplitude, the ellipse becomes a circle and this is said to be *circular polarization*. The antenna is said to be *circularly polarized*. When the phase difference is 0° or 180°, the ellipse becomes a line and this is referred to as *linear polarization*. The antenna is then *linearly polarized*.

2.3.1 Elliptical Polarization

We can write the electric far-field as

$$\mathbf{E} = E_\vartheta \hat{\mathbf{u}}_\vartheta + E_\varphi \hat{\mathbf{u}}_\varphi, \tag{2.26}$$

where E_ϑ is the complex amplitude of the ϑ-component of the electric field and E_φ is the complex amplitude of the φ-component of the electric field. The vectors $\hat{\mathbf{u}}_\vartheta$ and $\hat{\mathbf{u}}_\varphi$ are unit vectors in the ϑ- and φ-directions respectively. A brief overview of complex algebra and vector mathematics will be given in the next chapter.

The complex amplitudes E_ϑ and E_φ can be written as,

$$E_\vartheta = |E_\vartheta| e^{j\Psi_\vartheta}, \tag{2.27}$$

and

$$E_\varphi = |E_\varphi| e^{j\Psi_\varphi}, \tag{2.28}$$

where Ψ_ϑ and Ψ_φ represent the phases of E_ϑ and E_φ, respectively.

The electric field vector therefore may be written as

$$\mathbf{E} = |E_\vartheta| e^{j\Psi_\vartheta} \left(\hat{\mathbf{u}}_\vartheta + \rho \hat{\mathbf{u}}_\varphi \right), \tag{2.29}$$

where

$$\rho = \frac{|E_\varphi|}{|E_\vartheta|} e^{j\left(\Psi_\varphi - \Psi_\vartheta \right)}. \tag{2.30}$$

To trace the extremity of the electric field in the ϑ,φ-plane, the real part of the electric field needs to be taken

$$\mathbf{E} = \Re \left\{ \mathbf{E} e^{j2\pi ft} \right\}$$
$$= |E_\vartheta| \left\{ \cos \left(2\pi ft + \Psi_\vartheta \right) \hat{\mathbf{u}}_\vartheta + |\rho| \cos \left(2\pi ft + \Psi_\vartheta + \Psi_\rho \right) \hat{\mathbf{u}}_\varphi \right\}, \tag{2.31}$$

where f is the used frequency, t is the time and

$$\Psi \rho = \Psi_\varphi - \Psi_\vartheta. \tag{2.32}$$

Elimination of the time t results in a description of the trace of the extremity of the electric field vector in the ϑ,φ-plane. After some mathematical manipulations we get

$$\left(\frac{E_\vartheta}{|E_\vartheta|} \right)^2 + \left(\frac{E_\varphi}{|E_\varphi|} \right)^2 - \frac{2 E_\vartheta E_\varphi}{|E_\vartheta||E_\varphi|} \cos \left(\Psi_\rho \right) = \sin^2 \left(\Psi_\rho \right). \tag{2.33}$$

Dividing this equation by $\sin^2 \left(\Psi_\rho \right)$, results in

$$\left[\frac{1}{|E_\vartheta|^2 \sin^2 \left(\Psi_\rho \right)} \right] E_\vartheta^2 - \left[\frac{2 \cos \left(\Psi_\rho \right)}{|E_\vartheta||E_\varphi| \sin^2 \left(\Psi_\rho \right)} \right] E_\vartheta E_\varphi + \left[\frac{1}{|E_\varphi|^2 \sin^2 \left(\Psi_\rho \right)} \right] E_\varphi^2 = 1, \tag{2.34}$$

which is the equation of an ellipse in the ϑ,φ-plane, see Figure 2.23.

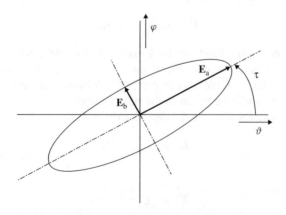

Figure 2.23 Polarization ellipse.

The tilt angle τ is given by

$$\tau = \arctan\left(\frac{|\mathbf{E}_\varphi|}{|\mathbf{E}_\vartheta|}\right). \tag{2.35}$$

The amount of ellipticity is expressed as the ratio of the semimajor and semiminor axes of the ellipse. This ratio is therefore known as the *axial ratio* and is usually expressed in decibels. With reference to Figure 2.23

$$AR = 20\log\left(\frac{|\mathbf{E}_a|}{|\mathbf{E}_b|}\right). \tag{2.36}$$

We will now show that circular polarization and linear polarization are special cases of elliptical polarization.

2.3.2 Circular Polarization

For the special situation where $|\mathbf{E}_\vartheta| = |\mathbf{E}_\varphi| = |\mathbf{E}|$ and $\Psi_\rho = \pm\frac{\pi}{2}$, equation (2.34) simplifies to

$$\left(\frac{\mathsf{E}_\vartheta}{|E|}\right)^2 + \left(\frac{\mathsf{E}_\varphi}{|E|}\right)^2 = 1, \tag{2.37}$$

which describes a circle in the ϑ,φ-plane.

We may further distinguish between the situations $\Psi_\rho = \frac{\pi}{2}$ and $\Psi_\rho = -\frac{\pi}{2}$. For the former situation we get, upon substitution in equation (2.31)

$$\mathsf{E}_L = |E|\left[\cos\left(2\pi f t'\right)\hat{\mathbf{u}}_\vartheta - \sin\left(2\pi f t'\right)\hat{\mathbf{u}}_\varphi\right], \tag{2.38}$$

where

$$2\pi f t' = 2\pi f t + \Psi_\rho. \tag{2.39}$$

Similarly, we get for the latter situation

$$\mathsf{E}_R = |E|\left[\cos\left(2\pi f t'\right)\hat{\mathbf{u}}_\vartheta + \sin\left(2\pi f t'\right)\hat{\mathbf{u}}_\varphi\right]. \tag{2.40}$$

For the first situation ($\Psi_\rho = \frac{\pi}{2}$) we have obtained an electric field vector that rotates counterclockwise with time, looking in the direction of propagation. This situation is depicted in Figure 2.22 and in Figure 2.24a. We call this circular polarization state *left-hand circular polarization* (LHCP), following the direction of a left-handed screw.

For the second situation ($\Psi_\rho = -\frac{\pi}{2}$) we have obtained an electric field vector that rotates clockwise with time, looking in the direction of propagation. This situation is depicted in Figure 2.24(b). We call this circular polarization state *right-hand circular polarization* (RHCP), following the direction of a right-handed screw.

2.3.3 Linear Polarization

For the special situation where $\Psi_\rho = \pm\pi$ and $|\mathbf{E}_\vartheta|$ not necessarily equal to $|\mathbf{E}_\varphi|$, equation (2.34) simplifies to

$$\frac{\mathsf{E}_\vartheta}{|E|} + \frac{\mathsf{E}_\varphi}{|E|} = 0, \tag{2.41}$$

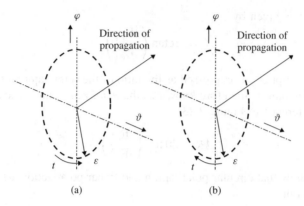

(a)   (b)

Figure 2.24 Circular polarization. (a) Left-hand circular polarization (LHCP). (b) Right-hand circular polarization (RHCP).

which describes a straight line in the ϑ,φ-plane, see also Figure 2.22(c). The *axial ratio* (AR) is infinite for a linearly polarized wave.

2.3.4   Axial Ratio

The axial ratio of a perfect circularly polarized wave is equal to unity. In practice, perfect circular polarization is seldom encountered; a certain amount of ellipticity will always exist. This ellipticity is expressed in the axial ratio, the ratio of the semimajor to the semiminor axis lengths of the polarization ellipse. These lengths may be expressed in terms of amplitudes of the LHCP and RHCP components. The elliptical polarization may be seen as consisting of a combination of LHCP and RHCP polarization, the dominant part of these two determining the direction of rotation of the elliptical polarization.

To show this we start by decomposing the electric field into LHCP and RHCP components:

$$\mathbf{E} = E_\vartheta \hat{\mathbf{u}}_\vartheta + E_\varphi \hat{\mathbf{u}}_\varphi = E_L \hat{\mathbf{u}}_L + E_R \hat{\mathbf{u}}_R, \qquad (2.42)$$

where

$$\hat{\mathbf{u}}_L = \frac{1}{\sqrt{2}} \left\{ \hat{\mathbf{u}}_\vartheta + j\hat{\mathbf{u}}_\varphi \right\}, \qquad (2.43)$$

$$\hat{\mathbf{u}}_R = \frac{1}{\sqrt{2}} \left\{ \hat{\mathbf{u}}_\vartheta - j\hat{\mathbf{u}}_\varphi \right\}, \qquad (2.44)$$

and

$$E_L = \frac{1}{\sqrt{2}} \left\{ E_\vartheta - jE_\varphi \right\}, \qquad (2.45)$$

$$E_R = \frac{1}{\sqrt{2}} \left\{ E_\vartheta + jE_\varphi \right\}. \qquad (2.46)$$

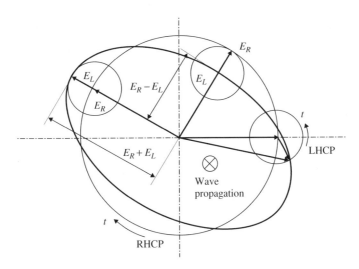

Figure 2.25 Right-hand elliptically polarized wave, decomposed into RHCP and LHCP components.

E_L is the LHCP component of the wave, E_R is the RHCP component of the wave. The relation between an elliptically polarized wave and the two circular polarization components is graphically represented in Figure 2.25.

In a predominantly RHCP wave (as shown in Figure 2.25), the RHCP component is called the *co-polarization* and the LHCP component is called the *cross-polarization*. The opposite is true for a predominantly LHCP wave.

The axial ratio is given by

$$AR = 20 \log \left| \frac{|E_L| + |E_R|}{|E_L| - |E_R|} \right|. \tag{2.47}$$

Circular polarization may be beneficial both in mobile satellite communications and in radar applications.

At L-band frequencies (1–2 GHz), the ionosphere acts as a so-called *Faraday rotator* [7], meaning that a linearly polarized wave undergoes a rotation upon passing through the ionosphere. The use of circularly polarized transmit and receive antennas will eliminate negative effects of this rotation (co-polarized signal attenuation and increase of cross-polarized signal level).

In radar, circular polarization may be employed to 'see through' rain [8]. A right-hand (left-hand) circularly polarized wave, incident upon a (near) spherical rain drop will be reflected as a left-hand (right-hand) circularly polarized wave, to which the receiving antenna is insensitive and thus this rain scatter will be rejected, while reflections from a complicated structure, such as an aircraft, will possess circularly polarized components with the right rotation direction to be accepted by the antenna. A linearly polarized wave, reflected from a raindrop, would be accepted by the antenna since the antenna will accept linearly polarized waves, 180° shifted in phase, equally as well as in-phase components.

2.4   Antenna Effective Area and Vector Effective Length

Any antenna, be it a horn antenna or even a half-wave (wire) dipole antenna, may be considered as an aperture antenna. This means that we may associate with every antenna an aperture or equivalent area that – in the case of a receiving antenna – extracts energy from an incident wave.

2.4.1   Effective Area

Let's assume an antenna in receive situation[5] with a plane wave incident upon it, having a power density at the position of the antenna of S (Wm^{-2}). We may characterize the antenna by a *maximum equivalent area* or *maximum equivalent aperture*, A_{em} that is defined as

$$A_{em} = \frac{P_T}{S},$$
(2.48)

where P_T is the available power at the terminals of the antenna.

The effective area is, strictly speaking, a direction-dependent quantity, but if no direction is specified – as in the above equation – the direction is assumed to be that of maximum directivity. Furthermore, without further specification, we assume that the polarization of the antenna and the impinging plane wave are lined up and that the antenna does not introduce dielectric or ohmic losses. Under these conditions the effective area as defined in the above equation is the *maximum effective area*.

Real aperture antennas, like electromagnetic horns, have effective apertures which are smaller than the physical ones. For electromagnetic horns, the effective aperture is in the order of 0.5 to 0.7 times the value of the physical aperture [9].

The maximum effective aperture of an antenna may be related to its directivity; this relation is derived in Appendix A and results in

$$D = \frac{4\pi A_{em}}{\lambda^2}.$$
(2.49)

We see that an increase in (effective) area leads to an increase in directivity. Antenna beamwidth is therefore inversely proportional to aperture size.

The *effective aperture*, A_e, is related to the *maximum effective aperture*, A_{em}, through the radiation efficiency, η, that accounts for ohmic and dielectric losses in the antenna

$$A_e = \eta A_{em}.$$
(2.50)

With equation (2.13) we then find the following relation between effective aperture and gain

$$G = \frac{4\pi A_e}{\lambda^2}.$$
(2.51)

---

[5] By virtue of reciprocity, which will be explained further on in this book, transmit and receive properties of antennas are identical.

Example

To show that the concept of effective area is a mathematical abstraction that is not necessarily related to physical area, we will calculate the effective area of a short dipole antenna of length $l \ll \lambda$ and negligible diameter [2, 3].

The antenna and its equivalent circuit are shown in Figure 2.26. For educational reasons we assume antenna and load impedances to be real. Furthermore, we assume the short dipole to be lossless (which is not true in practice!).

The open circuit voltage of the short dipole antenna is

$$V = |\mathbf{E}_i| l = E_i l, \tag{2.52}$$

where E_i is the amplitude of the incoming (lined up) linearly polarized electric field. The maximum available power, P_a, realized when $R_L = R_A$, is given by

$$P_a = \frac{|V|^2}{8R_A} = \frac{|E_i|^2 l^2}{8R_A}, \tag{2.53}$$

where the short dipole radiation resistance, R_A may be calculated as [2, 3, 9], see also Chapter 5.

$$R_A = 80\pi^2 \left(\frac{l}{\lambda}\right)^2, \tag{2.54}$$

where λ is the wavelength used.

The power density in the far-field region of the source, S, may be calculated as

$$S = \frac{1}{2}\left|\mathbf{E}_i \times \mathbf{H}_i^*\right| = \frac{1}{2}\frac{|E_i|^2}{Z_0}, \tag{2.55}$$

where Z_0 is the intrinsic impedance of free space, which is equal to $Z_0 = 120\pi$ (Ω).

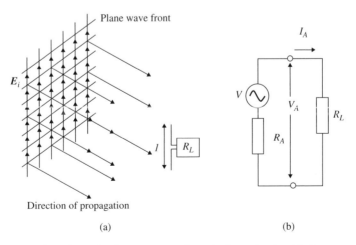

(a)   (b)

Figure 2.26 Receiving short dipole antenna and equivalent circuit. (a) Antenna connected to receiver with impedance R_L. (b) Equivalent circuit.

With equations (2.48), (2.53), (2.54) and (2.55), we then find for the maximum effective area, A_{em}, of a short dipole (that has a negligible physical area!)

$$A_{em} = \frac{P_a}{S} = \frac{3}{8\pi}\lambda^2. \tag{2.56}$$

An effective dimension associated with straight-wire antennas, that appeals more to our intuition is the *vector effective length*.

2.4.2 Vector Effective Length

The *vector effective length* or *vector effective height* is used to determine the open circuit voltage induced at the antenna terminals, when a plane wave is incident on the antenna. When polarizations of plane wave and antenna are lined up, the effective length of a short dipole antenna is identical to its physical length, l, see equation (2.52) in the example in Section 2.4.1.[6] In general, the effective length is a direction-dependent quantity [2].

$$\mathbf{l}_e(\vartheta, \varphi) = l_\vartheta(\vartheta, \varphi)\hat{\mathbf{u}}_\vartheta + l_\varphi(\vartheta, \varphi)\hat{\mathbf{u}}_\varphi. \tag{2.57}$$

The open circuit voltage of an antenna is then obtained by projecting the incident electric field vector on the complex conjugate of the vector effective length [3]

$$V_{oc} = \mathbf{E}_i \cdot \mathbf{l}_e^*, \tag{2.58}$$

where the complex conjugate is used to correct for the fact that the vector effective length is associated with the transmitting case, while the open circuit voltage is obtained in the receiving case.

The incident field vector – radiated by a transmitting antenna – may be expressed in terms of vector effective length according to [2, 10] as

$$\mathbf{E}_i = -jZ_0\frac{kI_{in}}{4\pi r}\mathbf{l}_e e^{-jkr}, \tag{2.59}$$

where I_{in} is the current at the input terminals of the transmitting antenna, r is the distance between transmitting and receiving antenna and $k = 2\pi/\lambda$.

Example
The radiated electric far-field of a vertically oriented half-wave dipole antenna is given by [3]

$$\mathbf{E}_i = jZ_0\frac{kI_{in}}{4\pi r}\left(\frac{\lambda}{\pi}\right)e^{-jkr}\sin(\vartheta)\frac{\cos\left[\left(\frac{\pi}{2}\right)\cos(\vartheta)\right]}{\sin^2(\vartheta)}\mathbf{u}_\vartheta. \tag{2.60}$$

Substitution of equation (2.60) for $\vartheta = \frac{\pi}{2}$ into equation (2.59) gives for the effective length of a half-wave dipole antenna

$$|\mathbf{l}_e| = l_\vartheta = \frac{\lambda}{\pi} = 0.32\lambda. \tag{2.61}$$

---

[6] For the short dipole in the example it was implicitly assumed that the current over the short dipole antenna is uniform. For any current distribution other than uniform, physical and effective length are not identical anymore.

The effective length of a half-wave dipole antenna is smaller than its physical length, just as we have seen that the effective aperture of an electromagnetic horn antenna is smaller than its physical size.

The concept of effective aperture or length is based on the principle that an imaginary aperture or wire antenna is conceived that intercepts the same amount of power as the original antenna does, but it does so *uniformly*, unlike the original antenna.[7] Only when the original antenna is very small, does the power interception approximately take place uniformly and effective and physical dimensions get close to one another. The concept of effective aperture is not restricted to physical aperture antennas and, likewise, the concept of effective length may also be applied to non-wire (aperture) antennas.

2.5   Radio Equation

As stated before, an antenna is never used as a stand-alone component, but will always be part of a communication or radar system. In a communication system, we have to deal with at least two antennas: a transmitting antenna and a receiving antenna. We assume that both antennas are lined up in terms of polarization and maximum directivity and that they are positioned in each other's far-field regions, see Figure 2.27.

The gain of the transmitting antenna is G_T, the gain of the receiving antenna is G_R. The power density S at distance R from the transmitting antenna is

$$S = G_T \frac{P_T}{4\pi R^2},$$
(2.62)

where P_T is the *input* power at the terminals of the transmitting antenna. The factor $1/(4\pi R^2)$ accounts for the spherical spreading of the energy. Since the transmit antenna is non-isotropic, this factor is multiplied by the gain of the antenna.

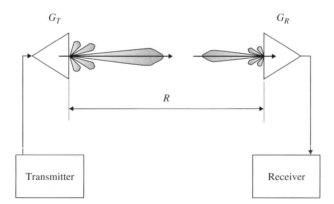

Figure 2.27   Communication system consisting of a transmitter having an antenna with gain G_T and a receiver having an antenna with gain G_R. The antennas are displaced a distance R and assumed to be in each other's far-field regions and lined up with respect to polarization and directivity.

---

[7] We have assumed that the half-wave dipole has a sinusoidal current distribution over the wire. For the determination of the effective length, a constant current is assumed. Therefore, the effective length (0.32λ) is shorter than the physical length (0.50λ).

The amount of power intercepted by the receiving antenna, P_R, is this power density multiplied by the effective area of the receive antenna, A_{eR}. This effective area is directly related to the gain of the receive antenna,

$$P_R = SA_{eR} = G_T \frac{P_T}{4\pi R^2} G_R \frac{\lambda^2}{4\pi}. \qquad (2.63)$$

After rearranging the terms of this equation, we may relate the power received at the terminals of the receive antenna, P_R, to the power delivered at the terminals of the transmit antenna, P_T, as

$$\frac{P_R}{P_T} = \left(\frac{\lambda}{4\pi R}\right)^2 G_R G_T. \qquad (2.64)$$

This equation is known as the *radio equation* or the *Friis transmission equation*. The term $(\lambda/(4\pi R))^2$ is known as the *free space loss factor*.

Example

Assume a broadcasting system, operating at $100\,\text{MHz}$, employing a half-wave dipole antenna, having a gain of $2.15\,\text{dBi}$.[8] The power accepted by the transmit antenna is $1\,\text{kW}$.

The minimum required power delivered by the receiving antenna is $1\,\text{nW}$. When the maximum range is $500\,\text{km}$, what should be the minimum gain of the receiving antenna?

The required gain is given by

$$G_R = \frac{16\pi^2 R^2 P_R}{\lambda^2 P_T G_T}. \qquad (2.65)$$

The wavelength is given by $\lambda = c_0/f$, where c_0 is the velocity of light in vacuum $(c_0 \approx 3 \cdot 10^8\,\text{ms}^{-1})$ and f is the frequency. The wavelength therefore is $3.0\,\text{m}$. The gain of the transmitting antenna is $G_T = 10^{2.15/10} = 1.64$.[9] The desired minimum gain is then found to be

$$G_R = \frac{16 \cdot 9.87 \cdot 2.5 \cdot 10^{11} \cdot 10^{-9}}{9 \cdot 10^3 \cdot 1.64} = 2.67, \qquad (2.66)$$

or $G_R = 10\log(2.67) = 4.27\,\text{dB}_i$.

Mind that the equation is solved for the situation where both antennas are lined up with respect to gain and polarization.

Example

Consider a mobile communication system consisting of two identical transmitter-receiver sets operating at $1\,\text{GHz}$. The same half-wave dipole antenna is used both for transmission and for reception. $G_T = G_R = G = 2.15\,\text{dBi}$. The power delivered to the antenna in transmission is $1\,\text{W}$. The minimum power at the antenna terminals in reception is $-65\,\text{dBm}$. Find the maximum allowable distance, R, between the two sets.

---

[8] The units of dBi refer to a gain relative to an isotropic radiator. Another unit sometimes encountered is the dBd, the gain relative to an elementary dipole radiator ($2.15\,\text{dBi} = 0.0\,\text{dBd}$).

[9] The product $P_T G_T$ is known as the *effective isotropic radiated power* (EIRP). The EIRP is the power intensity that could have been obtained from an isotropic radiator if it had an input power, P_{in} equal to the EIRP.

The maximum allowable distance is given by

$$R = \frac{\lambda}{4\pi} G \sqrt{\frac{P_T}{P_R}}. \tag{2.67}$$

The wavelength is $c_0/f = 3 \cdot 10^8/10^9 = 0.30$ m. $G = 1.64$. $P_T = 1$ W. The minimum power at reception is expressed in dBm, meaning decibels with respect to 1 mW. Therefore, $P_R = 10^{-65/10}$ mW $= 10^{-65/10} \cdot 10^{-3}$ W $= 3.16 \cdot 10^{-10}$ W. The maximum distance is then found to be

$$R = \frac{0.30}{12.57} 1.64 \sqrt{\frac{1}{3.16 \cdot 10^{-10}}} = 2.20 \text{ km}. \tag{2.68}$$

2.6 Radar Equation

A radar system is very much like the communication system described in the previous section. The difference is that instead of a direct link, the electromagnetic waves, emitted by the transmitter T in Figure 2.28(a), now reach the receiver, R, via a reflection against a target.

The power density incident on the target, S_i, is given by

$$S_i = G_T \frac{P_T}{4\pi R_1^2}, \tag{2.69}$$

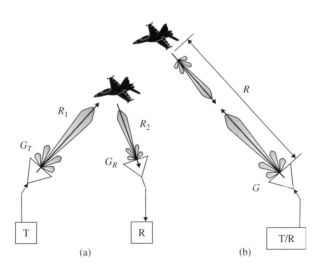

(a)   (b)

Figure 2.28 Radar system (a) Bistatic radar. The transmitter is connected to an antenna with gain G_T and is at a distance R_1 from the reflecting target. The receiver has an antenna with gain G_R and is at a distance R_2 from the target. (b) Monostatic radar. The same antenna, having gain G, is used for transmission and reception. The distance to the target is R. The target is assumed to be in each antenna's far-field region.

where G_T is the gain of the transmit antenna, P_T is the power at the input of the transmit antenna and R_1 is the distance between transmit antenna and target. It is implicitly assumed that the beam of the transmit antenna is directed to the target.

The power intercepted by the target, P_i, is proportional to the power density, S_i and is given by

$$P_i = \sigma S_i, \tag{2.70}$$

where σ (m^2) is known as the *radar cross-section* (RCS) of the target.

2.6.1 Radar Cross-Section

The radar cross-section of a target is the equivalent area intercepting that amount of power that, when scattered equally in all directions, produces an echo at the radar equal to that coming from the target [8].

There is in general not a simple rule, relating the physical size of a target to its radar cross-section, although, in general, larger targets exhibit larger RCSs. Table 2.1 [2, 8] gives an overview of some typical RCS values.

We now return to the radar equation.

The power density at the position of the receiver after scattering from the target, S_s is given by

$$S_s = \frac{P_i}{4\pi R_2^2} = \frac{\sigma G_T P_T}{(4\pi)^2 R_1^2 R_2^2}. \tag{2.71}$$

Table 2.1 Typical RCS values

Object	RCS (m^2)
Conventional, winged missile	0.5
Small fighter or 4-passenger jet	2
Large fighter	6
Medium bomber or medium jet airliner	20
Large bomber or large jet airliner	40
Jumbo jet	100
Small pleasure boat	2
Cabin cruiser	10
Pickup truck	200
Automobile	100
Bicycle	2
Man	1
Bird	0.01
Insect	0.00001

The power available at the receiver, P_R, is

$$P_R = A_{er} S_s, \tag{2.72}$$

where A_{er} is the effective aperture of the receiving antenna. With use of equation (2.51), we find

$$P_R = P_T \frac{\sigma G_T G_R \lambda^2}{(4\pi)^3 R_1^2 R_2^2}. \tag{2.73}$$

This equation, relating received and transmitted power, is known as the *radar equation*.

The setup with a separate transmit and receive antenna, physically displaced, see Figure 2.28(a), is called *bistatic radar*. The more common *monostatic radar*, see Figure 2.28(b), uses the same antenna both for transmission and reception. The consequences for the radar equation are that $G_T = G_R = G$ and $R_1 = R_2 = R$, leading to

$$P_R = P_T \frac{\sigma G^2 \lambda^2}{(4\pi)^3 R^4}. \tag{2.74}$$

Example

It is necessary to detect a target with a RCS of 1m^2 at a range of 150 km. A monostatic radar is used. The gain of the antenna employed is 40 dB at a frequency of 3 GHz. The minimum power at the terminals of the antenna in receiving mode is -100 dBm.

What is the transmitting power needed and what is the size of the effective antenna aperture?

The transmitting power is given by

$$P_T = P_R \frac{(4\pi)^3 R^4}{\sigma G^2 \lambda^2}. \tag{2.75}$$

$G = 10^{40/10} = 10^4$ and $\lambda = 3 \cdot 10^8/3 \cdot 10^9 = 10^{-1}$ m. $P_R = 10^{-100/10}$ mW $= 10^{-13}$ W. The transmitted power is then found to be

$$P_T = 10^{-13} \frac{1984.40 \left(150 \cdot 10^3\right)^4}{10^8 \cdot 10^{-2}} \text{W} = 100.46 \text{ kW}. \tag{2.76}$$

The effective aperture is

$$A_e = \frac{G \lambda^2}{4\pi} = \frac{10^4 \cdot 10^{-2}}{12.57} \text{ m}^2 = 7.96 \text{ m}^2. \tag{2.77}$$

Example

Answer the same questions if the frequency is now 1 GHz. We assume that the antenna is replaced by an antenna that has a gain of 40 dBi at 1 GHz.

The wavelength has changed to $\lambda = 0.30$ m. The transmitted power is now found to be

$$P_T = 10^{-13} \frac{1984.40 \left(150 \cdot 10^3\right)^4}{10^8 \cdot 0.30^2} \text{W} = 11.16 \text{ kW}. \tag{2.78}$$

The effective aperture of the antenna is

$$A_e = \frac{G\lambda^2}{4\pi} = \frac{10^4 \cdot 0.30^2}{12.57}\,\mathrm{m^2} = 71.60\,\mathrm{m^2}. \tag{2.79}$$

Thus, at a lower frequency we may detect targets using less power, but we need larger antennas.

2.7 Problems

2.1 What is meant with the 'far field' of an antenna?

2.2 (a) A rectangular aperture has dimensions 30 cm × 15 cm and is operated at a frequency of 10 GHz. At what distance does the far field start?

   (b) A parabolic (dish) reflector has a diameter $D = 80$ cm and a focal distance $F = 13$ cm. A circular aperture (horn antenna) is placed at the focal distance. We assume that the reflector is positioned in the far-field of the horn antenna. What is the maximum diameter of this horn antenna if the frequency is 1.5 GHz?

2.3 In Figure 2.29 the normalized radiation pattern cut of a certain antenna is shown. Determine, from the figure, the half power beamwidth of this cut.

2.4 (a) The radiation efficiency of an antenna is 0.95 and the gain is 3.0 dB. What is the directivity?

   (b) Is the gain always higher or always lower than the directivity, and why?

2.5 (a) The Mars Rover that landed on the planet Mars in 2003 is equipped with an X-band (8–12 GHz) high-gain antenna. The gain is 25 dBi at 8.4 GHz. The minimum

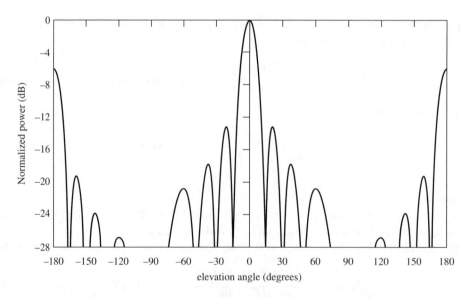

Figure 2.29 Normalized radiation pattern cut of a certain antenna.

required power level at reception is −156 dBm. The distance to Earth is 56 million kilometers. What is the EIRP needed on Earth to transmit directly to the Rover?

(b) If the effective aperture size of the transmit antenna is 1 m², what is the transmitted power?

2.6 A GSM-900 (900 MHz) base station receiver employs an antenna with a gain of 18 dBi and has a sensitivity of −50 dBm. When a GSM phone is transmitting at 0.1 W employing an antenna having a gain of 2.15 dBi, what is the maximum reachable, line-of-sight distance between phone and base station?

2.7 A monostatic radar is used to detect an unknown flying object. From time measurements, the distance is found to be 100 km. The transmit power is 100 kW, the antenna gain is 30 dBi. The received power is measured to be 10^{-13} W. What is the radar cross section of the unknown object and what might it be?

2.8 We want to use a radar to detect birds at a distance up to 100 m. We will use an antenna having a gain of 18 dBi and a receiver having a sensitivity of −80 dBm at 10 GHz. How much power must we use?

References

1. Hubregt J. Visser, *Array and Phased Array Antenna Basics*, John Wiley & Sons, Chichester, UK, 2005.
2. Constantine E. Balanis, *Antenna Theory, Analysis and Design*, *second edition*, John Wiley & Sons, New York, 1997.
3. Warren L. Stutzman and Gary A. Thiele, *Antenna Theory and Design*, John Wiley & Sons, New York, 1998.
4. R.C. Johnson (ed.) *Antenna Engineering Handbook*, *third edition*, McGraw-Hill, New York, 1993.
5. Roger F. Harrington, *Time-Harmonic Electromagnetic Fields*, John Wiley & Sons, New York, 2001.
6. David M. Pozar, *Microwave Engineering*, *second edition*, John Wiley & Sons, New York, 1998.
7. G. Murakami and G.S. Wickizer, Ionosphere Phase Distortion and Faraday Rotation of Radio Waves, *RCA Review*, Vol. 30, 1969, pp. 488–491.
8. Merill I. Skolnik, *Introduction to Radar Systems*, McGraw-Hill, Auckland, 1981.
9. John D. Kraus, *Antennas*, McGraw-Hill, New York, 1950.
10. R.E. Collin and F.J. Zucker, *Antenna Theory, Part I*, McGraw-Hill, New York, 1969.

3

Vector Analysis

Scalars are defined by an amplitude only, for example temperature, charge. Vectors, for example forces, are defined not only by an amplitude but also by a direction. This means that simple arithmetic operations that can be applied for scalars, like addition, subtraction, multiplication and differentiation, have slightly more complicated counterparts for vectors. In this chapter we will briefly explain these operations and we will discuss the *gradient*-, *divergence*- and *curl*-operators.

A *scalar* is represented in print as a normal, though often italic, letter (e.g., charge q). A vector is represented in print with a bold face letter (e.g., electric field **E**). Since bold face letters are difficult to reproduce in writing, the convention of writing a small arrow over the letter is adapted. This arrow may be simplified to a half arrow or a line. Also a line underneath the letter may be used to represent a vector. Thus, the vector **a** may be represented as

$$\mathbf{a} = \vec{a} = \bar{a} = \underline{a}. \tag{3.1}$$

The amplitude of this vector, $|\mathbf{a}|$, is a scalar, $|\mathbf{a}| = a$.

For a vector **a** in a Cartesian three-dimensional space, the relation between amplitude and direction (ϑ, φ), see Figure 3.1, is given by

$$\mathbf{a} = a \sin(\vartheta) \cos(\varphi)\hat{\mathbf{u}}_x + a \sin(\vartheta) \sin(\varphi)\hat{\mathbf{u}}_y + a \cos(\vartheta)\hat{\mathbf{u}}_z. \tag{3.2}$$

Here, $\hat{\mathbf{u}}_x$, $\hat{\mathbf{u}}_y$ and $\hat{\mathbf{u}}_z$ are unit vectors in the x-, y- and z-directions, respectively. Unit vectors have a length 1 and are mutually orthogonal. Customarily they are indicated with a 'hat'.

Having clarified notational issues, we are now ready for discussing vector analysis.

3.1 Addition and Subtraction

We start with the addition of two vectors **a** and **b**. Therefore we construct a parallelogram from the vectors **a** and **b** as shown in Figure 3.2 and find the sum of the two vectors $\mathbf{a} + \mathbf{b}$ as the vector starting at the origin and following the long diagonal of the parallelogram.

Antenna Theory and Applications, First Edition. Hubregt J. Visser.
© 2012 John Wiley & Sons, Ltd. Published 2012 by John Wiley & Sons, Ltd.

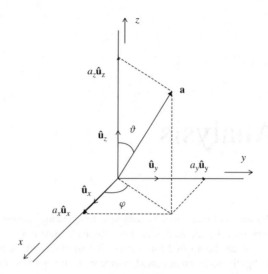

Figure 3.1 Three-dimensional vector **a** having amplitude a and direction (ϑ, φ).

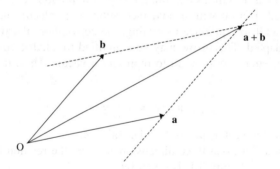

Figure 3.2 Addition of vectors **a** and **b**.

For the subtraction $\mathbf{a} - \mathbf{b}$ we apply the same method. First we obtain the vector $-\mathbf{b}$ as the image of **b** upon reflection through the origin. Then, $\mathbf{a} - \mathbf{b} = \mathbf{a} + (-\mathbf{b})$, see Figure 3.3.

3.2   Products

For the scalar quantities a and b, only the product $a \cdot b$ exists. For vectors, however, we have two possibilities: The so-called *scalar product* or *dot product* an the *vector product* or *cross product*.

3.2.1   Scalar Product or Dot Product

The dot product of two vectors results in a scalar. The result thus only has an amplitude and no direction. The dot product of two vectors **a** and **b** is obtained by the orthogonal projection of one of the vectors onto the other one, see Figure 3.4.

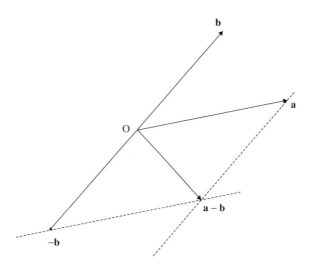

Figure 3.3 Subtraction of vectors **a** and **b**.

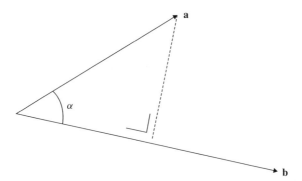

Figure 3.4 Obtaining the dot product of the vectors **a** and **b** that are positioned with respect to each other at an angle α.

The dot product **a** · **b** is

$$\mathbf{a} \cdot \mathbf{b} = |\mathbf{a}|.|\mathbf{b}|. \cos(\alpha) = a.b. \cos(\alpha). \tag{3.3}$$

Note that for $\alpha = \pi/2$ the dot product is 0. It does not matter which vector is projected onto which vector, the dot product is commutative:

$$\mathbf{a} \cdot \mathbf{b} = \mathbf{b} \cdot \mathbf{a}. \tag{3.4}$$

3.2.2 Vector Product or Cross Product

The result of the cross product of two vectors **a** and **b** is a vector again, thus having amplitude *and* direction. The direction is that of a right-hand screw, see Figure 3.5.

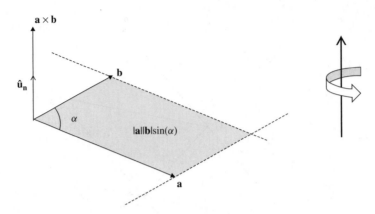

Figure 3.5 Obtaining the cross product of the vectors **a** and **b** that are positioned with respect to each other at an angle α.

The amplitude is equal to the surface of the parallelogram constructed from the two vectors **a** and **b**.

The cross product $\mathbf{a} \times \mathbf{b}$ is determined by

$$\mathbf{a} \times \mathbf{b} = |\mathbf{a}| . |\mathbf{b}| . \sin(\alpha)\hat{\mathbf{u}}_n, \tag{3.5}$$

where $\hat{\mathbf{u}}_n$ is the so-called *normal* (length *1*) on the surface defined by **a** and **b**. In contrast to the dot product, the order of vectors in the cross product does matter, the cross product is *not* commutative:

$$\mathbf{a} \times \mathbf{b} = -\mathbf{b} \times \mathbf{a}. \tag{3.6}$$

3.2.3 Triple Product

With the dot product and the cross product defined in the previous sections we now have two possibilities for forming a triple product, $(\mathbf{a} \times \mathbf{b}) \cdot \mathbf{c}$ and $(\mathbf{a} \times \mathbf{b}) \times \mathbf{c}$.

3.2.3.1   (a x b) · c

We start with determining the cross product of the vectors **a** and **b**. Thereto, we decompose the vectors into their Cartesian components:

$$\begin{aligned} \mathbf{a} &= a_x\hat{\mathbf{u}}_x + a_y\hat{\mathbf{u}}_y + a_z\hat{\mathbf{u}}_z, \\ \mathbf{b} &= b_x\hat{\mathbf{u}}_x + b_y\hat{\mathbf{u}}_y + b_z\hat{\mathbf{u}}_z, \end{aligned} \tag{3.7}$$

and apply the determinant representation for determining the cross product:

$$(\mathbf{a} \times \mathbf{b}) = \begin{vmatrix} \hat{\mathbf{u}}_x & \hat{\mathbf{u}}_y & \hat{\mathbf{u}}_z \\ a_x & a_y & a_z \\ b_x & b_y & b_z \end{vmatrix} = (a_yb_z - b_ya_z)\hat{\mathbf{u}}_x - (a_xb_z - b_xa_z)\hat{\mathbf{u}}_y + (a_xb_y - b_xa_y)\hat{\mathbf{u}}_z. \tag{3.8}$$

Next we take the dot product with $\mathbf{c} = c_x\hat{\mathbf{u}}_x + c_y\hat{\mathbf{u}}_y + c_z\hat{\mathbf{u}}_z$ and finally obtain

$$(\mathbf{a} \times \mathbf{b}) \cdot \mathbf{c} = (a_yb_z - b_ya_z)c_x - (a_xb_z - b_xa_z)c_y + (a_xb_y - b_xa_y)c_z. \tag{3.9}$$

3.2.3.2   (a × b) × c

Using equation (3.8) and the determinant representation we obtain through laborious but straightforward calculations and a regrouping of common terms:

$$(\mathbf{a} \times \mathbf{b}) \times \mathbf{c} = \begin{vmatrix} \hat{\mathbf{u}}_x & \hat{\mathbf{u}}_y & \hat{\mathbf{u}}_z \\ a_yb_z - b_ya_z & a_zb_x - b_za_x & a_xb_y - b_xa_y \\ c_x & c_y & c_z \end{vmatrix}$$

$$= \left\{(a_zb_x - b_za_x)c_z - (a_xb_y - b_xa_y)c_y\right\} \hat{\mathbf{u}}_x$$
$$- \left\{(a_yb_z - b_ya_z)c_z - (a_xb_y - b_xa_y)c_x\right\} \hat{\mathbf{u}}_y$$
$$+ \left\{(a_yb_z - b_ya_z)c_y - (a_zb_x - b_za_x)c_x\right\} \hat{\mathbf{u}}_z$$

$$= (\mathbf{a} \cdot \mathbf{c})\mathbf{b} - (\mathbf{b} \cdot \mathbf{c})\mathbf{a}. \tag{3.10}$$

3.3   Differentiation

Suppose that a vector $\mathbf{a}$ is a function of the (scalar) parameter t:

$$\mathbf{a} = \mathbf{a}(t). \tag{3.11}$$

For a small change δt, the vector changes into, see Figure 3.6,

$$\mathbf{a}(t + \delta t) = \mathbf{a} + \delta\mathbf{a}. \tag{3.12}$$

We now define the derivative of vector $\mathbf{a}$ with respect to t in the usual way

$$\frac{d\mathbf{a}}{dt} = \lim_{\delta t \to 0} \frac{\delta\mathbf{a}}{\delta t}. \tag{3.13}$$

The derivative with respect to t of the product of two scalar functions $a(t)$ and $b(t)$ is found with the aid of the chain rule

$$\frac{d}{dt}(ab) = \frac{da}{dt}b + a\frac{db}{dt}. \tag{3.14}$$

To apply the chain rule to the dot product of two vectors $\mathbf{a}(t)$ and $\mathbf{b}(t)$, we first write the dot product as in equation (3.3) $\mathbf{a} \cdot \mathbf{b} = ab\cos(\alpha)$, where α is the angle between the two vectors. Application of the chain rule then results in

$$\frac{d}{dt}(\mathbf{a} \cdot \mathbf{b}) = \frac{da}{dt}b\cos(\alpha) + a\frac{db}{dt}\cos(\alpha). \tag{3.15}$$

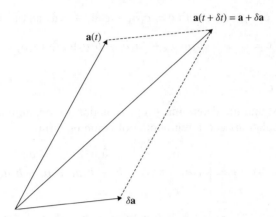

Figure 3.6   Change of vector **a**(t) into vector **a**$(t + \delta t) = \mathbf{a} + \delta\mathbf{a}$.

As δt approaches zero, the limit of $\delta\mathbf{a}$ and the limit of $\delta\mathbf{b}$ will also approach zero, and therefore $\frac{d\mathbf{a}}{dt}$ and $\frac{d\mathbf{b}}{dt}$ will be directed along **a** and **b**, respectively. Equation (3.15) then becomes

$$\frac{d}{dt}(\mathbf{a} \cdot \mathbf{b}) = \frac{d\mathbf{a}}{dt} \cdot \mathbf{b} + \mathbf{a} \cdot \frac{d\mathbf{b}}{dt}. \tag{3.16}$$

In a similar way we find that

$$\frac{d}{dt}(\mathbf{a} \times \mathbf{b}) = \frac{d\mathbf{a}}{dt} \times \mathbf{b} + \mathbf{a} \times \frac{d\mathbf{b}}{dt}. \tag{3.17}$$

In discussing the product of two vectors, we found that it was necessary to introduce two types of product. Now, it is necessary to introduce three types of differentiation. These types are: *gradient*, *divergence* and *curl*.

3.3.1 Gradient

The gradient works on a scalar quantity (e.g., temperature) and identifies the greatest rate of increase of this scalar quantity. The gradient therefore has an amplitude *and* direction. The gradient is a vector quantity of a scalar field.

The gradient of the scalar ϕ, in Cartesian coordinates, is given by

$$grad(\phi) = \frac{\delta\phi}{\delta x}\hat{\mathbf{u}}_x + \frac{\delta\phi}{\delta y}\hat{\mathbf{u}}_y + \frac{\delta\phi}{\delta z}\hat{\mathbf{u}}_z. \tag{3.18}$$

To determine the spatial variation of a scalar function in an arbitrarily chosen direction, it suffices to determine the component of the gradient in that direction.

For a shorthand notation we will make use of a so-called *operator notation*. To start with, we write equation (3.18) as

$$grad(\phi) = \left(\frac{\delta}{\delta x}\hat{\mathbf{u}}_x + \frac{\delta}{\delta y}\hat{\mathbf{u}}_y + \frac{\delta}{\delta z}\hat{\mathbf{u}}_z\right)\phi. \tag{3.19}$$

It is important to understand that in equation (3.19) we have *NOT* written a multiplication. Instead, the equation means that the *operation* indicated between the parentheses acts upon ϕ. How it acts on ϕ is defined in equation (3.18).

Next, we write the operator, that is the term between parentheses in equation (3.19), in shorthand

$$\nabla \equiv \frac{\delta}{\delta x}\hat{\mathbf{u}}_x + \frac{\delta}{\delta y}\hat{\mathbf{u}}_y + \frac{\delta}{\delta z}\hat{\mathbf{u}}_z, \tag{3.20}$$

and use this to write equation (3.18) in shorthand, using the *nabla-operator*, as

$$grad(\phi) = \nabla\phi, \tag{3.21}$$

where the nabla-operator is given by equation (3.20) and the operation is defined by equation (3.18).

3.3.2 Divergence

The divergence (a scalar quantity) is the outward directed flux, per unit of volume, acting on a point at a vector field.

To explain this, let's assume a vector field $\mathbf{a}$ and a point P in this field. Next, consider a small volume δV around P. The (closed) surface of this volume we call S. The total flux that leaves the volume is given by

$$\iint_S \mathbf{a} \cdot d\mathbf{S}, \tag{3.22}$$

where $d\mathbf{S}$ is perpendicular to S, see Figure 3.7.

The flux per unit of volume is now given by

$$\frac{\iint_S \mathbf{a} \cdot d\mathbf{S}}{\delta V}. \tag{3.23}$$

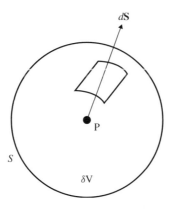

Figure 3.7   Volume δV having closed surface S. $\delta \mathbf{S}$ is perpendicular to S.

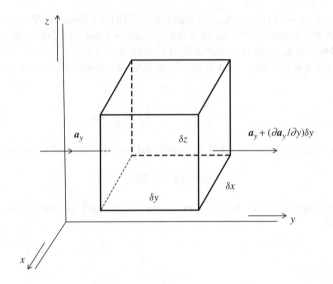

Figure 3.8 Rectangular volume with sides δx, δy and δz.

From this equation follows the definition of the divergence in the point P

$$div(\mathbf{a}) = \lim_{\delta V \to 0} \frac{\oiint_S \mathbf{a} \cdot d\mathbf{S}}{\delta V}. \tag{3.24}$$

To get a more practical equation, we consider – in a Cartesian coordinate system – a rectangular volume having sides δx, δy and δz, see Figure 3.8.

The net flux in the y-direction, see also Figure 3.8, is now given by

$$\left(a_y + \frac{\partial a_y}{\partial y}\delta y\right)\delta x \delta z - a_y \delta x \delta z = \frac{\partial a_y}{\partial y}\delta x \delta y \delta z. \tag{3.25}$$

In a similar way we find for the net flux in the x-direction and for the net flux in the z-direction

$$\left(a_x + \frac{\partial a_x}{\partial x}\delta x\right)\delta y \delta z - a_x \delta y \delta z = \frac{\partial a_x}{\partial x}\delta x \delta y \delta z, \tag{3.26}$$

$$\left(a_z + \frac{\partial a_z}{\partial z}\delta z\right)\delta x \delta y - a_z \delta x \delta y = \frac{\partial a_z}{\partial z}\delta x \delta y \delta z. \tag{3.27}$$

Then, the net flux out of the volume is given by

$$\left(\frac{\partial a_x}{\partial x} + \frac{\partial a_y}{\partial y} + \frac{\partial a_z}{\partial z}\right)\delta x \delta y \delta z, \tag{3.28}$$

and by applying equation (3.24), the divergence is given by

$$div(\mathbf{a}) = \lim_{\delta x \delta y \delta z \to 0} \frac{\left(\frac{\partial a_x}{\partial x} + \frac{\partial a_y}{\partial y} + \frac{\partial a_z}{\partial z}\right)\delta x \delta y \delta z}{\delta x \delta y \delta z}, \tag{3.29}$$

or

$$div(\mathbf{a}) = \frac{\partial a_x}{\partial x} + \frac{\partial a_y}{\partial y} + \frac{\partial a_z}{\partial z}. \tag{3.30}$$

This equation may also be written in shorthand using the operator notation. Again we use the nabla-operator, see equation (3.20), to obtain

$$\nabla \cdot \mathbf{a} = \left(\frac{\partial}{\partial x}\hat{\mathbf{u}}_x + \frac{\partial}{\partial y}\hat{\mathbf{u}}_y + \frac{\partial}{\partial z}\hat{\mathbf{u}}_z \right) \cdot (a_x\hat{\mathbf{u}}_x + a_y\hat{\mathbf{u}}_y + a_z\hat{\mathbf{u}}_z) = \frac{\partial a_x}{\partial x} + \frac{\partial a_y}{\partial y} + \frac{\partial a_z}{\partial z}, \tag{3.31}$$

and so

$$div(\mathbf{a}) = \nabla \cdot \mathbf{a}. \tag{3.32}$$

3.3.2.1 Gradient and Divergence Calculus

Now that the gradient and divergence have been explained, it is useful to derive some calculation rules involving the gradient and divergence. We need to keep in mind that the gradient operates on a scalar and results in a vector and that the divergence operates on a vector and results in a scalar.

We start by repeating the operator notation

$$\begin{aligned} grad(\phi) &= \nabla\phi \\ div(\mathbf{a}) &= \nabla \cdot \mathbf{a}, \end{aligned} \tag{3.33}$$

where

$$\nabla \equiv \frac{\partial}{\partial x}\hat{\mathbf{u}}_x + \frac{\partial}{\partial y}\hat{\mathbf{u}}_y + \frac{\partial}{\partial z}\hat{\mathbf{u}}_z. \tag{3.34}$$

Now we can calculate the *divergence* of the product of a scalar ϕ and a vector $\mathbf{a}$

$$\begin{aligned} \nabla \cdot (\phi\mathbf{a}) &= \left(\frac{\partial}{\partial x}\hat{\mathbf{u}}_x + \frac{\partial}{\partial y}\hat{\mathbf{u}}_y + \frac{\partial}{\partial z}\hat{\mathbf{u}}_z \right) \cdot (\phi a_x\hat{\mathbf{u}}_x + \phi a_y\hat{\mathbf{u}}_y + \phi a_z\hat{\mathbf{u}}_z) \\ &= \frac{\partial(\phi a_x)}{\partial x} + \frac{\partial(\phi a_y)}{\partial y} + \frac{\partial(\phi a_z)}{\partial z} \\ &= a_x\frac{\partial\phi}{\partial x} + \phi\frac{\partial a_x}{\partial x} + a_y\frac{\partial\phi}{\partial y} + \phi\frac{\partial a_y}{\partial y} + a_z\frac{\partial\phi}{\partial z} + \phi\frac{\partial a_z}{\partial z} \\ &= \mathbf{a} \cdot \nabla\phi + \phi\nabla \cdot \mathbf{a}, \end{aligned} \tag{3.35}$$

or

$$div(\phi\mathbf{a}) = \mathbf{a} \cdot grad(\phi) + \phi div(\mathbf{a}). \tag{3.36}$$

3.3.3 Curl

The curl of a vector may be regarded as the generalization of current density. This statement needs an explanation.

Let us consider a thin wire, carrying a current **I**, see Figure 3.9.

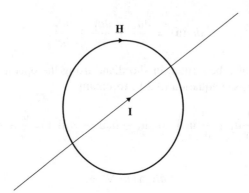

Figure 3.9 A current **I** induces a magnetic field **H**.

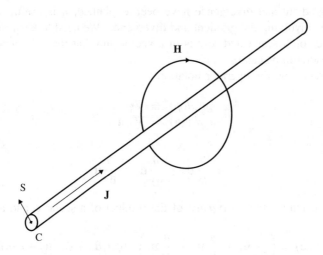

Figure 3.10 A current density **J** induces a magnetic field **H**.

The magnetic field, induced by the current, satisfies

$$\oint \mathbf{H} \cdot d\mathbf{l} = \mathbf{I}, \tag{3.37}$$

where $d\mathbf{l}$ is the integration variable that is tangent to the loop around the current. We now generalize to a current density **J** on a wire with cross section **S**, see Figure 3.10.

$$\oint \mathbf{H} \cdot d\mathbf{l} = \iint_S \mathbf{J} \cdot d\mathbf{S}. \tag{3.38}$$

We define the curl through equation (3.38), a generalization of current density,

$$\oint \mathbf{a} \cdot d\mathbf{l} = \iint_S curl(\mathbf{a}) \cdot d\mathbf{S}. \tag{3.39}$$

From this equation follows the definition of the curl

$$curl(\mathbf{a}) = \lim_{\delta S \to 0} \frac{\oint \mathbf{a} \cdot d\mathbf{l}}{\delta S}, \tag{3.40}$$

where δS is an infinitesimal part of the surface S. The curl is perpendicular to the contour C. The contour integral in the right-hand part of equation (3.40) explains the name of the operation: *curl*.

Just as with the definition of the divergence, equation (3.24),[1] this equation is not very well suited for practical applications. For a more practical formulation we consider, in a Cartesian coordinate system, a rectangular contour in the xz-plane, having sides δx and δz, see Figure 3.11. A vector $\mathbf{a}$ starting from the origin of the coordinate system has components a_x, a_y and a_z.

Following the contour, starting at the origin of the coordinate system in the direction of the block arrows, we encounter the following contributions to the contour integral

$$a_z \delta z$$
$$\left(a_x + \frac{\partial a_x}{\partial z} \delta z \right) \delta x$$
$$- \left(a_z + \frac{\partial a_z}{\partial x} \delta x \right) \delta z$$
$$- a_x \delta x \tag{3.41}$$

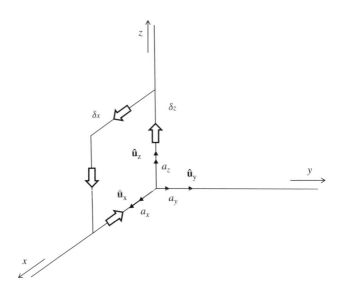

Figure 3.11   A rectangular contour in the xz-plane having sides δx and δz.

---

[1] Just as the curl may be considered to be a generalization of current density, the divergence may be considered as a generalization of charge density.

The integral over the contour shown in Figure 3.11 thus amounts to

$$\oint \mathbf{a} \cdot d\mathbf{l} = \left(\frac{\partial a_x}{\partial z} - \frac{\partial a_z}{\partial x} \right) \delta x \delta z, \tag{3.42}$$

and thus the y-component of the curl of vector $\mathbf{a}$ is

$$(curl(\mathbf{a}))_y = \frac{\partial a_x}{\partial z} - \frac{\partial a_z}{\partial x}. \tag{3.43}$$

In a similar way, by constructing contours in the yz-plane and in the x,y-plane, we find the x- and z-components, respectively, of the curl of vector $\mathbf{a}$:

$$(curl(\mathbf{a}))_x = \frac{\partial a_z}{\partial y} - \frac{\partial a_y}{\partial z}, \tag{3.44}$$

$$(curl(\mathbf{a}))_z = \frac{\partial a_y}{\partial x} - \frac{\partial a_x}{\partial y}. \tag{3.45}$$

The curl of vector $\mathbf{a}$ may be written, in operator notation, as

$$curl(\mathbf{a}) = \begin{vmatrix} \hat{\mathbf{u}}_x & \hat{\mathbf{u}}_y & \hat{\mathbf{u}}_z \\ \frac{\partial}{\partial x} & \frac{\partial}{\partial y} & \frac{\partial}{\partial z} \\ a_x & a_y & a_z \end{vmatrix} = \left(\frac{\partial}{\partial x}\hat{\mathbf{u}}_x + \frac{\partial}{\partial y}\hat{\mathbf{u}}_y + \frac{\partial}{\partial z}\hat{\mathbf{u}}_z \right) \times \mathbf{a}. \tag{3.46}$$

In equation (3.46) we recognize the nabla-operator, equation (3.20), so that we may write the curl of vector $\mathbf{a}$ in shorthand as

$$curl(\mathbf{a}) = \nabla \times \mathbf{a}. \tag{3.47}$$

3.3.3.1 Gradient, Divergence and Curl Calculus

Now that, next to the gradient and the divergence, the curl has also been explained, we may derive calculation rules for cascaded operations. We repeat that the gradient operates on a scalar and results in a vector, that the divergence operates on a vector and results in a scalar and that the curl operates on a vector and results in a vector.

The expressions for cascaded operations are found by consequently applying

$$\nabla \equiv \frac{\partial}{\partial x}\hat{\mathbf{u}}_x + \frac{\partial}{\partial y}\hat{\mathbf{u}}_y + \frac{\partial}{\partial z}\hat{\mathbf{u}}_z$$

$$grad(\phi) = \nabla\phi = \frac{\partial\phi}{\partial x}\hat{\mathbf{u}}_x + \frac{\partial\phi}{\partial y}\hat{\mathbf{u}}_y + \frac{\partial\phi}{\partial z}\hat{\mathbf{u}}_z$$

$$div(\mathbf{a}) = \nabla \cdot \mathbf{a} = \frac{\partial a_x}{\partial x} + \frac{\partial a_y}{\partial y} + \frac{\partial a_z}{\partial z} \tag{3.48}$$

$$curl(\mathbf{a}) = \nabla \times \mathbf{a} = \begin{vmatrix} \hat{\mathbf{u}}_x & \hat{\mathbf{u}}_y & \hat{\mathbf{u}}_z \\ \frac{\partial}{\partial x} & \frac{\partial}{\partial y} & \frac{\partial}{\partial z} \\ a_x & a_y & a_z \end{vmatrix}$$

Using equations (3.48), we find

$$div\,(curl(\mathbf{a})) = \nabla \cdot \nabla \times \mathbf{a} = 0, \tag{3.49}$$

$$curl\,(grad(\phi)) = \nabla \times \nabla\phi = 0, \tag{3.50}$$

$$curl\,(curl(\mathbf{a})) = \nabla \times \nabla \times \mathbf{a} = \nabla(\nabla \cdot \mathbf{a}) - \nabla^2\mathbf{a}, \tag{3.51}$$

$$curl(\phi\mathbf{a}) = \nabla \times (\phi\mathbf{a}) = (\nabla\phi) \times \mathbf{a} + \phi(\nabla \times \mathbf{a}) = grad(\phi) \times \mathbf{a} + \phi curl(\mathbf{a}). \tag{3.52}$$

Useful vector formulas are given in Appendix B.

3.4 Problems

3.1 Calculate the dot product $\mathbf{a} \cdot \mathbf{b}$ if
   (a) $\mathbf{a} = \hat{\mathbf{u}}_x + 2\hat{\mathbf{u}}_y$ and $\mathbf{b} = 3\hat{\mathbf{u}}_x + 4\hat{\mathbf{u}}_y$,
   (b) $\mathbf{a} = 3\hat{\mathbf{u}}_x + 5\hat{\mathbf{u}}_y$ and $\mathbf{b} = -7\hat{\mathbf{u}}_x + 12\hat{\mathbf{u}}_y$,
   (c) $\mathbf{a} = -\hat{\mathbf{u}}_x - \hat{\mathbf{u}}_y$ and $\mathbf{b} = -\hat{\mathbf{u}}_x + \hat{\mathbf{u}}_y$.

3.2 Prove that $|\mathbf{a}| = |\mathbf{b}|$, if $(\mathbf{a} + \mathbf{b})$ and $(\mathbf{a} - \mathbf{b})$ are perpendicular.

3.3 Find the angle between the vectors $\mathbf{a}$ and $\mathbf{b}$ if
   (a) $\mathbf{a} = \hat{\mathbf{u}}_x + \sqrt{3}\hat{\mathbf{u}}_y$ and $\mathbf{b} = \sqrt{3}\hat{\mathbf{u}}_x + \hat{\mathbf{u}}_y$,
   (b) $\mathbf{a} = 3\hat{\mathbf{u}}_x + 7\hat{\mathbf{u}}_y$ and $\mathbf{b} = -2\hat{\mathbf{u}}_x + 5\hat{\mathbf{u}}_y$,
   (c) $\mathbf{a} = \hat{\mathbf{u}}_x + 2\hat{\mathbf{u}}_y$ and $\mathbf{b} = 3\hat{\mathbf{u}}_x + \hat{\mathbf{u}}_y$.

3.4 Calculate the cross product $\mathbf{a} \times \mathbf{b}$ if
   (a) $\mathbf{a} = 4\hat{\mathbf{u}}_x - 4\hat{\mathbf{u}}_y - 3\hat{\mathbf{u}}_z$ and $\mathbf{b} = \hat{\mathbf{u}}_x$,
   (b) $\mathbf{a} = 4\hat{\mathbf{u}}_x - 4\hat{\mathbf{u}}_y - 3\hat{\mathbf{u}}_z$ and $\mathbf{b} = \hat{\mathbf{u}}_x - \hat{\mathbf{u}}_y$,
   (c) $\mathbf{a} = \hat{\mathbf{u}}_x + \hat{\mathbf{u}}_y - 4\hat{\mathbf{u}}_z$ and $\mathbf{b} = 2\hat{\mathbf{u}}_x - \hat{\mathbf{u}}_y + 3\hat{\mathbf{u}}_z$.

3.5 If $\phi = 2xy^2 + x^2z^3$, what is $\nabla\phi$?

3.6 If $\phi = x^2y + 2xy$, what is the gradient at the point $(x, y, z) = (2, 1, 1)$?

3.7 Find the gradient of $f(x, y, z) = xyz$.

3.8 Find the gradient of $f(x, y, z) = (x - 1)^2 - y^2 + (z + 3)^2$.

3.9 If $\mathbf{f}(x, y, z) = xyz\hat{\mathbf{u}}_x + 3y^2z\hat{\mathbf{u}}_y - 2x^2yz\hat{\mathbf{u}}_z$, find the divergence of $\mathbf{f}(x, y, z)$.

3.10 Find the divergence of $\mathbf{f} = 2x^2z\hat{\mathbf{u}}_x - 3xz^2\hat{\mathbf{u}}_y + xy^2z\hat{\mathbf{u}}_z$.

3.11 Find the divergence of $\mathbf{f} = x\hat{\mathbf{u}}_x + y\hat{\mathbf{u}}_y + z\hat{\mathbf{u}}_z$.

3.12 If $\mathbf{r} = x\hat{\mathbf{u}}_x + y\hat{\mathbf{u}}_y + z\hat{\mathbf{u}}_z$, find the divergence of $\mathbf{f} = \frac{\mathbf{r}}{|\mathbf{r}|^3}$.

3.13 Calculate the curl of $\mathbf{f}(x, y, z) = 3x^2y\hat{\mathbf{u}}_x + 2xy^2z\hat{\mathbf{u}}_y - xyz^3\hat{\mathbf{u}}_z$.

3.14 Calculate the curl of the gradient of ϕ, if $\phi = 4xy^2z^3$.

3.15 Calculate the curl of $\mathbf{f}(x, y, z) = x^2z^3\hat{\mathbf{u}}_x + x^3yz\hat{\mathbf{u}}_y - y^2z^4\hat{\mathbf{u}}_z$.

4

Radiated Fields

To be able to calculate the radiated fields of an antenna, we start by calculating the radiated fields of an arbitrary current density. In the next chapter we will make an assessment of this current density for a specific antenna, that is the thin-wire dipole antenna. Then, the radiated fields will be a special case of the general situation described in this chapter. The origin of the calculations is the Maxwell equations, wherein current density and charge density are the sources. By introducing the magnetic vector potential and the Lorentz gauge, we will be able to calculate the radiated fields, based on the current density only.

4.1 Maxwell Equations

The Maxwell equations, in the time-domain, are given by:

$$\nabla \times \boldsymbol{E}(\mathbf{r}, t) = -\frac{\partial \boldsymbol{B}(\mathbf{r}, t)}{\partial t}, \tag{4.1}$$

$$\nabla \times \boldsymbol{H}(\mathbf{r}, t) = \frac{\partial \boldsymbol{D}(\mathbf{r}, t)}{\partial t} + \boldsymbol{J}_e(\mathbf{r}, t), \tag{4.2}$$

$$\nabla \cdot \boldsymbol{B}(\mathbf{r}, t) = 0, \tag{4.3}$$

$$\nabla \cdot \boldsymbol{D}(\mathbf{r}, t) = \rho_e(\mathbf{r}, t). \tag{4.4}$$

In these equations, $\boldsymbol{E}$ is the electric field $\left[Vm^{-1}\right]$, $\boldsymbol{D}$ is the dielectric displacement $\left[Cm^{-2}\right]$, $\boldsymbol{B}$ is the magnetic induction $\left[Wbm^{-2}\right]$, $\boldsymbol{H}$ is the magnetic field $\left[Am^{-1}\right]$, ρ_e is the electric charge density $\left[Cm^{-3}\right]$ and $\boldsymbol{J}_e$ is the electric current density $\left[Am^{-2}\right]$.

For electromagnetic waves propagating in free space

$$\boldsymbol{D}(\mathbf{r}, t) = \varepsilon_0 \boldsymbol{E}(\mathbf{r}, t), \tag{4.5}$$

$$\boldsymbol{B}(\mathbf{r}, t) = \mu_0 \boldsymbol{H}(\mathbf{r}, t), \tag{4.6}$$

where ε_0 is the free space permittivity, $\varepsilon_0 \approx 8.854 \times 10^{-12}$ $\left[Fm^{-1}\right]$ and μ_0 is the free space permeability, $\mu_0 = 4\pi \times 10^{-7}$ $\left[Hm^{-1}\right]$.

Antenna Theory and Applications, First Edition. Hubregt J. Visser.
© 2012 John Wiley & Sons, Ltd. Published 2012 by John Wiley & Sons, Ltd.

In electromagnetics we work with time-harmonic fields, meaning that

$$E(\mathbf{r}, t) = \Re\left\{\mathbf{E}(\mathbf{r})e^{j\omega t}\right\}, \tag{4.7}$$

$$H(\mathbf{r}, t) = \Re\left\{\mathbf{H}(\mathbf{r})e^{j\omega t}\right\}, \tag{4.8}$$

$$B(\mathbf{r}, t) = \Re\left\{\mathbf{B}(\mathbf{r})e^{j\omega t}\right\}, \tag{4.9}$$

$$D(\mathbf{r}, t) = \Re\left\{\mathbf{D}(\mathbf{r})e^{j\omega t}\right\}, \tag{4.10}$$

where ω is the angular frequency, $\omega = 2\pi f$, f being the frequency. For complex algebra, see Appendix C.

Substituting equations (4.7)–(4.10) and equations (4.5) and (4.6) in equations (4.1)–(4.4) results in

$$\nabla \times \mathbf{E}(\mathbf{r}) = -j\omega\mu_0 \mathbf{H}(\mathbf{r}), \tag{4.11}$$

$$\nabla \times \mathbf{H}(\mathbf{r}) = j\omega\varepsilon_0 \mathbf{E}(\mathbf{r}) + \mathbf{J}_e(\mathbf{r}), \tag{4.12}$$

$$\nabla \cdot \mathbf{H}(\mathbf{r}) = 0, \tag{4.13}$$

$$\nabla \cdot \mathbf{E}(\mathbf{r}) = \frac{\rho_e(\mathbf{r})}{\varepsilon_0}. \tag{4.14}$$

The general problem now is to find the complex fields $\mathbf{E}(\mathbf{r})$ and $\mathbf{H}(\mathbf{r})$, given the sources $\mathbf{J}_e(\mathbf{r})$ and $\rho_e(\mathbf{r})$. Once these complex fields are found, the physical fields are obtained through equations (4.7) and (4.8).

If the source distributions are known, the complex field may be obtained through integrating, that is adding contributions, over the source volume. For this method it is necessary, however, to know both the electric current density $\mathbf{J}_e(\mathbf{r})$ and the electric charge density $\rho_e(\mathbf{r})$. By using a so-called *vector potential*, we will be able to relate the electric current density to the electric charge density and then calculate the radiated fields using the electric current density only. In general, it will be possible to assess the electric current density, as we will see in this and the subsequent chapters. For the electric charge density an assessment is not always obvious.

4.2 Vector Potential

From equation (4.13), $\nabla \cdot \mathbf{H} = 0$,[1] and the identity (3.49), $\nabla \cdot (\nabla \times \mathbf{A}_e) = 0$, it follows that we may write the magnetic field as

$$\mathbf{H} = \frac{1}{\mu_0} \nabla \times \mathbf{A}_e, \tag{4.15}$$

where $\mathbf{A}_e$ is called the *magnetic vector potential*.

Substituting equation (4.15) in equation (4.11) gives

$$\nabla \times \mathbf{E} = -j\omega\nabla \times \mathbf{A}_e - \nabla \times \nabla\phi_e. \tag{4.16}$$

---

[1] From here on we omit writing down the **r**-dependences explicitly.

The addition of the last term in the right-hand-side of equation (4.16) is allowed since, from equation (3.50), $\nabla \times \nabla \phi_e = 0$. ϕ_e is an arbitrary electric scalar potential. The electric field may now be written as

$$\mathbf{E} = -j\omega\mathbf{A}_e - \nabla\phi_e. \tag{4.17}$$

Substitution of equations (4.15) and (4.17) in equation (4.12) results in

$$\nabla \times \nabla \times \mathbf{A}_e = k_0^2\mathbf{A}_e - j\omega\varepsilon_0\mu_0\nabla\phi_e + \mu_0\mathbf{J}_e, \tag{4.18}$$

where $k_0 = \omega\sqrt{\varepsilon_0\mu_0}$ is the free-space wave number.

Using the vector identity (3.51), $\nabla \times \nabla \times \mathbf{A}_e = \nabla (\nabla \cdot \mathbf{A}_e) - \nabla^2\mathbf{A}_e$, the equation becomes

$$\nabla^2\mathbf{A}_e + k_0^2\mathbf{A}_e = \nabla (\nabla \cdot \mathbf{A}_e) + j\omega\varepsilon_0\mu_0\nabla\phi_e - \mu_0\mathbf{J}_e. \tag{4.19}$$

In this equation we may already start to recognize a Helmholtz equation. We may simplify this equation by defining the scalar potential ϕ_e, which is allowed since we introduced this scalar potential as being arbitrary. By choosing $j\omega\varepsilon_0\mu_0\phi_e = -\nabla \cdot \mathbf{A}_e$ or

$$\phi_e = -\frac{1}{j\omega\varepsilon_0\mu_0}\nabla \cdot \mathbf{A}_e, \tag{4.20}$$

equation (4.19) reduces to

$$\nabla^2\mathbf{A}_e + k_0^2\mathbf{A}_e = -\mu_0\mathbf{J}_e, \tag{4.21}$$

a vectorial, inhomogeneous Helmholtz equation, relating the source $\mathbf{J}_e$ to the vector potential $\mathbf{A}_e$. The particular choice for the scalar potential in equation (4.20) is known as the *Lorentz gauge* or *Lorentz condition*.

The source ρ_e may be related to the scalar potential ϕ_e by substituting equation (4.17) in equation (4.14):

$$\nabla \cdot \mathbf{E} = -j\omega\nabla \cdot \mathbf{A}_e - \nabla^2\phi_e, \tag{4.22}$$

and applying the Lorentz gauge $\nabla \cdot \mathbf{A}_e = -j\omega\varepsilon_0\mu_0\phi_e$, leading to

$$\nabla^2\phi_e + k_0^2\phi_e = -\frac{\rho_e}{\varepsilon_0}. \tag{4.23}$$

This is a scalar, inhomogeneous Helmholtz equation relating the source ρ_e to the scalar potential ϕ_e.

We now want to obtain a solution for the vector potential in equation (4.21). To simplify the process, we start with a z-directed constant current of infinitesimal length l at the origin of a rectangular coordinate system, see Figure 4.1.

With the current density now given by $\mathbf{J}_e = I_0 l\delta(\mathbf{r})\hat{\mathbf{u}}_z$, the vectorial, inhomogeneous Helmholtz equation results in a scalar one:

$$\nabla^2 A_{e_z} + k_0^2 A_{e_z} = -\mu_0 I_0 l\delta(\mathbf{r}). \tag{4.24}$$

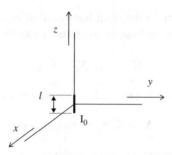

Figure 4.1 z-directed constant current I_0 of infinitesimal length l at the origin of a rectangular coordinate system.

To solve for A_{e_z} we consider two situations.

1. *Static situation* In the static situation, $\omega = 0$ and thus $k_0 = 0$. Equation (4.24) then reduces to Poisson's equation

$$\nabla^2 A_{e_z} = -\mu_0 I_0 l \delta(r). \tag{4.25}$$

Knowing that Poisson's equation for the electrostatic potential, from equation (4.23) for $k_0 = 0$, has solution [1, 2]

$$\phi_e = \frac{\rho_e}{4\pi \varepsilon_0 r}, \tag{4.26}$$

we find as a solution for equation (4.24)

$$A_{e_z} = \frac{\mu_0 I_0 l \delta(r)}{4\pi r}. \tag{4.27}$$

2. *Outside the source volume* Outside the source volume the current density is zero and the Helmholtz equation (4.24) reduces to a homogeneous one

$$\nabla^2 A_{e_z} + k_0^2 A_{e_z} = 0. \tag{4.28}$$

The solution of this equation is of the form

$$A_{e_z} = C \frac{e^{-jk_0 r}}{r}, \tag{4.29}$$

where C is a constant still to be determined.[2] If we look at the static situation, that is $k_0 = 0$, equation (4.28) reduces to the Laplace equation

$$\nabla^2 A_{e_z} = 0, \tag{4.30}$$

---

[2] A solution to equation (4.28) could also be of the form $C \exp(+jk_0 r)/r$. However, we have implicitly assumed a time-dependence according to $\exp(j\omega t)$, meaning that the phase of the solution would be $(+jk_0 r + j\omega t)$. To see the movement of the phase front, we need to keep this phase constant for increasing t. We see that we can only accomplish this for decreasing r. Therefore this solution creates waves moving towards the source. Following the same reasoning we can see that for the chosen solution, equation (4.29), we have created waves moving away from the source.

having a solution of the form[3]

$$A_{e_z} = \frac{C}{r}. \tag{4.31}$$

By comparing equations (4.29) and (4.31) we see that the only difference between a static and a time-varying solution is a multiplicative factor e^{-jk_0r}.

The constant C may be obtained through comparing the solution for the time-varying situation, equation (4.29), with the solution for the static situation, equation (4.27), to obtain $C = \mu_0 I_0 l \delta(r)/4\pi$.

$$A_{e_z} = \frac{\mu_0 I_0 l \delta(r)}{4\pi r} e^{-jk_0r}. \tag{4.32}$$

We will now generalize this result. We start with the orientation of the infinitesimal dipole.

The analysis we have performed for a z-directed constant current may be applied to all Cartesian components of an arbitrarily oriented, infinitesimal current at the origin. Combining all the vector potential components then leads to

$$\mathbf{A}_e = \frac{\mu_0 I_0 l \delta(r)}{4\pi r} e^{-jk_0r} \hat{\mathbf{u}}_r, \tag{4.33}$$

where $\hat{\mathbf{u}}_r = \mathbf{r}/r$.

Next, we move the infinitesimal dipole out of the origin and place it at a position $\mathbf{r}_0 = x_0 \hat{\mathbf{u}}_x + y_0 \hat{\mathbf{u}}_y + z_0 \hat{\mathbf{u}}_z$, see Figure 4.2(a).

Equation (4.33) then becomes

$$\mathbf{A}_e = I_0 l (\mathbf{r}_0) \frac{\mu_0}{4\pi} \frac{e^{-jk_0 |\mathbf{r}-\mathbf{r}_0|}}{|\mathbf{r} - \mathbf{r}_0|} = I_0 l (\mathbf{r}_0) G (\mathbf{r}, \mathbf{r}_0), \tag{4.34}$$

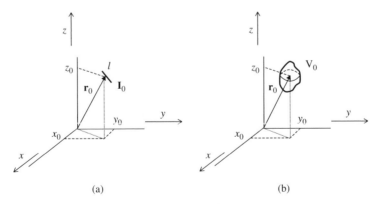

(a)   (b)

Figure 4.2   Generalization of the vector potential. (a) Infinitesimal current density placed at position $\mathbf{r}_0$. (b) Infinitesimal current density replaced by volume current density.

---

[3] Since A_{e_z} is a function of r only, we write the Laplace equation in spherical coordinates to obtain $\frac{d^2 A_{e_z}}{dr^2} + \frac{2}{r} \frac{dA_{e_z}}{dr} = 0$. Then it can be easily seen that a solution must be of the form $\frac{C}{r}$.

where

$$|\mathbf{r} - \mathbf{r}_0| = \sqrt{(x - x_0)^2 + (y - y_0)^2 + (z - z_0)^2}. \qquad (4.35)$$

$G(\mathbf{r}, \mathbf{r}_0)$ is called the *Green's function* of this problem.[4] It describes the response (vector potential) at observation position $\mathbf{r}$, due to a point source (current density) of unit strength at source position $\mathbf{r}_0$.

Finally, we abandon the infinitesimal dipole of strength $I_0 l$ and assume that we have a current density in a source volume V_0, see Figure 4.2(b). We now add all the contributions at all the positions $\mathbf{r}_0$ within V_0 to obtain

$$\mathbf{A}_e = \frac{\mu_0}{4\pi} \iiint_{V_0} \mathbf{J}_e(\mathbf{r}_0) \frac{e^{-jk_0|\mathbf{r}-\mathbf{r}_0|}}{|\mathbf{r} - \mathbf{r}_0|} dV_0 = \iiint_{V_0} \mathbf{J}_e(\mathbf{r}_0) G(\mathbf{r}, \mathbf{r}_0) dV_0. \qquad (4.36)$$

Similarly, we find for the generalized scalar potential

$$\phi_e = \frac{1}{4\pi\varepsilon_0} \iiint_{V_0} \rho_e(\mathbf{r}_0) \frac{e^{-jk_0|\mathbf{r}-\mathbf{r}_0|}}{|\mathbf{r} - \mathbf{r}_0|} dV_0 = \frac{1}{\mu_0\varepsilon_0} \iiint_{V_0} \rho_e(\mathbf{r}_0) G(\mathbf{r}, \mathbf{r}_0) dV_0. \qquad (4.37)$$

Having reached this point, it should be remembered that we introduced the concept of the vector potential to be able to calculate the radiated fields based on the current density only. From equation (4.36) we see that it suffices to know the current density only to calculate the vector potential. We have seen, see equation (4.15), that we can then calculate the radiated magnetic field from the vector potential

$$\mathbf{H} = \frac{1}{\mu_0} \nabla \times \mathbf{A}_e. \qquad (4.38)$$

The electric field is calculated as, see equation (4.17)

$$\mathbf{E} = -j\omega\mathbf{A}_e - \nabla\phi_e. \qquad (4.39)$$

By substituting the Lorentz gauge, equation (4.20), we obtain

$$\mathbf{E} = -j\omega\mathbf{A}_e - \nabla\left(-\frac{\nabla \cdot \mathbf{A}_e}{j\omega\varepsilon_0\mu_0}\right) = -j\omega\mathbf{A}_e + \frac{\nabla\nabla \cdot \mathbf{A}_e}{j\omega\varepsilon_0\mu_0}. \qquad (4.40)$$

Then, by using the vector identity (3.51), $\nabla \times \nabla \times \mathbf{A}_e = \nabla\nabla \cdot \mathbf{A}_e - \nabla^2\mathbf{A}_e$, we get

$$\mathbf{E} = -j\omega\mathbf{A}_e + \frac{\nabla^2\mathbf{A}_e}{j\omega\varepsilon_0\mu_0} + \frac{\nabla \times \nabla \times \mathbf{A}_e}{j\omega\varepsilon_0\mu_0} = \frac{1}{j\omega\varepsilon_0\mu_0}\left[\nabla \times \nabla \times \mathbf{A}_e + \nabla^2\mathbf{A}_e + k_0^2\mathbf{A}_e\right]. \qquad (4.41)$$

Substituting equation (4.21), $\nabla^2\mathbf{A}_e + k_0^2\mathbf{A}_e = -\mu_0\mathbf{J}_e$, in equation (4.41) finally results in

$$\mathbf{E} = \frac{1}{j\omega\varepsilon_0\mu_0}\left[\nabla \times \nabla \times \mathbf{A}_e - \mu_0\mathbf{J}_e\right]. \qquad (4.42)$$

---

[4] The Green's function of this problem is also seen as being defined excluding the term $\frac{\mu_0}{4\pi}$.

So, with equations (4.38) and (4.42) we are able to calculate the radiated fields everywhere, using the current density only.

Most of the time, we do not want to calculate the radiated fields everywhere, but only at large distances from the antenna, that is in the far-field, well outside the source volume V_0. Outside the source volume, $\mathbf{J}_e = 0$ and the magnetic and electric fields are calculated using

$$\mathbf{H} = \frac{1}{\mu_0} \nabla \times \mathbf{A}_e,$$

$$\mathbf{E} = \frac{1}{j\omega\varepsilon_0\mu_0} \nabla \times \nabla \times \mathbf{A}_e, \tag{4.43}$$

$$\mathbf{A}_e(\mathbf{r}) = \frac{\mu_0}{4\pi} \iiint_{V_0} \mathbf{J}_e \frac{e^{-jk_0|\mathbf{r}-\mathbf{r}_0|}}{|\mathbf{r}-\mathbf{r}_0|} dV_0.$$

4.3 Far-Field Approximations

Antennas will be used for communication over distances that are much larger than the antenna or the wavelength used. Therefore we need to evaluate the radiated fields at large distances. For this so-called far-field we may introduce approximations. These approximations will be based on the fact that the distances between the observation point P at $\mathbf{r}$ and source points at $\mathbf{r}_0$, $|\mathbf{r}_d| = |\mathbf{r}_0 - \mathbf{r}|$, will approximate those between the observation point and the origin for $r = |\mathbf{r}|$ becoming very large, see Figure 4.3.

4.3.1 Magnetic Field

The starting point is equation (4.43):

$$\mathbf{H} = \frac{1}{\mu_0} \nabla \times \mathbf{A}_e, \tag{4.44}$$

where

$$\mathbf{A}_e = \iiint_{V_0} \mathbf{J}_e(\mathbf{r}_0) G(\mathbf{r}, \mathbf{r}_0) dV_0, \tag{4.45}$$

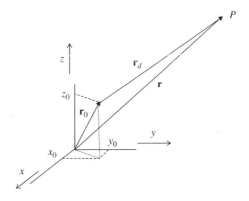

Figure 4.3 $|\mathbf{r}_d| = |\mathbf{r}_0 - \mathbf{r}|$ will approximate $r = |\mathbf{r}|$ for r becoming very large.

and

$$G(\mathbf{r}, \mathbf{r}_0) = \frac{\mu_0}{4\pi} \frac{e^{-jk_0|\mathbf{r}-\mathbf{r}_0|}}{|\mathbf{r} - \mathbf{r}_0|}. \tag{4.46}$$

The magnetic field may thus be written as

$$\mathbf{H} = \frac{1}{\mu_0} \iiint_{V_0} \nabla_r \times \mathbf{J}_e(\mathbf{r}_0) G(\mathbf{r}, \mathbf{r}_0) dV_0, \tag{4.47}$$

where we have used the subscripted nabla-operator ∇_r to stress the fact that the operation is on the observation point.

Using equation (3.52) for the curl of the product of a scalar and a vector we may write, for the kernel of the integration

$$\nabla_r \times \mathbf{J}_e(\mathbf{r}_0) G(\mathbf{r}, \mathbf{r}_0) = G(\mathbf{r}, \mathbf{r}_0) \nabla_r \times \mathbf{J}_e(\mathbf{r}_0) + [\nabla_r G(\mathbf{r}, \mathbf{r}_0)] \times \mathbf{J}_e(\mathbf{r}_0)$$

$$= [\nabla_r G(\mathbf{r}, \mathbf{r}_0)] \times \mathbf{J}_e(\mathbf{r}_0) \tag{4.48}$$

In the above, use has been made of $\nabla_r \times \mathbf{J}_e(\mathbf{r}_0) = 0$ since the nabla-operator operates on the observation point $(\mathbf{r})$ and not on the source point $(\mathbf{r}_0)$.

Equation (4.48) may be expanded using the chain rule for the derivative applied to equation (4.46).[5]

$$\nabla_r G(\mathbf{r}, \mathbf{r}_0) = \frac{\mu_0}{4\pi} \nabla_r \left[\frac{1}{|\mathbf{r} - \mathbf{r}_0|} \right] e^{-jk_0|\mathbf{r}-\mathbf{r}_0|} + \frac{\mu_0}{4\pi} \frac{1}{|\mathbf{r} - \mathbf{r}_0|} \nabla_r \left[e^{-jk_0|\mathbf{r}-\mathbf{r}_0|} \right]$$

$$= \frac{\mu_0}{4\pi} \left[-\frac{\mathbf{r} - \mathbf{r}_0}{|\mathbf{r} - \mathbf{r}_0|^3} \right] e^{-jk_0|\mathbf{r}-\mathbf{r}_0|} + \frac{\mu_0}{4\pi} \frac{1}{|\mathbf{r} - \mathbf{r}_0|} \left[-jk_0 e^{-jk_0|\mathbf{r}-\mathbf{r}_0|} \frac{\mathbf{r} - \mathbf{r}_0}{|\mathbf{r} - \mathbf{r}_0|} \right]$$

$$= -\frac{\mu_0}{4\pi} \frac{\mathbf{r} - \mathbf{r}_0}{|\mathbf{r} - \mathbf{r}_0|} e^{-jk_0|\mathbf{r}-\mathbf{r}_0|} \left[\frac{1}{|\mathbf{r} - \mathbf{r}_0|^2} + j\frac{k_0}{|\mathbf{r} - \mathbf{r}_0|} \right] \tag{4.49}$$

Since $\mathbf{r} - \mathbf{r}_0 = \mathbf{r}_d$, see Figure 4.3, and $r_d = |\mathbf{r}_d|$, we can write equation (4.49) as

$$\nabla_r G(\mathbf{r}, \mathbf{r}_0) = -\frac{\mu_0}{4\pi} \frac{e^{-jk_0 r_d}}{r_d} \left(jk_0 + \frac{1}{r_d} \right) \hat{\mathbf{u}}_{r_d}, \tag{4.50}$$

where

$$\hat{\mathbf{u}}_{r_d} = \frac{\mathbf{r}_d}{r_d}. \tag{4.51}$$

If we substitute equation (4.50) in equation (4.47), we find for the magnetic field

$$\mathbf{H} = -\frac{1}{4\pi} \iiint_{V_0} \left(jk_0 + \frac{1}{r_d} \right) \frac{e^{-jk_0 r_d}}{r_d} \hat{\mathbf{u}}_{r_d} \times \mathbf{J}_e(\mathbf{r}_0) dV_0. \tag{4.52}$$

---

[5] Use could be made of equation (3.20), $\nabla \equiv \frac{\delta}{\delta x}\hat{\mathbf{u}}_x + \frac{\delta}{\delta y}\hat{\mathbf{u}}_y + \frac{\delta}{\delta z}\hat{\mathbf{u}}_z$, and $|\mathbf{r} - \mathbf{r}_0| = \sqrt{(x - x_0)^2 + (y - y_0)^2 + (z - z_0)^2}$.

This equation will be used to apply to far-field approximations. To start with, we rewrite the equation for r_d in such a way that we may easily apply a Taylor series expansion [3]:[6]

$$
\begin{aligned}
r_d &= \sqrt{(x - x_0)^2 + (y - y_0)^2 + (z - z_0)^2} \\
&= \left\{ \left(x^2 + y^2 + z^2 \right) - 2\left(xx_0 + yy_0 + zz_0 \right) + \left(x_0^2 + y_0^2 + z_0^2 \right) \right\}^{\frac{1}{2}} \\
&= \left\{ r^2 - 2\left(xx_0 + yy_0 + zz_0 \right) + \left(x_0^2 + y_0^2 + z_0^2 \right) \right\}^{\frac{1}{2}} \\
&= r \left\{ 1 - \frac{2}{r^2}\left(xx_0 + yy_0 + zz_0 \right) + \frac{x_0^2 + y_0^2 + z_0^2}{r^2} \right\}^{\frac{1}{2}}
\end{aligned}
\tag{4.53}
$$

Applying a Taylor expansion now to r_d results in

$$
\begin{aligned}
r_d = r \left\{ 1 + \frac{1}{2}\left[-\frac{2}{r^2}\left(xx_0 + yy_0 + zz_0 \right) + \frac{x_0^2 + y_0^2 + z_0^2}{r^2} \right] \right. \\
\left. - \frac{1}{8}\left[-\frac{2}{r^2}\left(xx_0 + yy_0 + zz_0 \right) + \frac{x_0^2 + y_0^2 + z_0^2}{r^2} \right]^2 + \cdots \right\}.
\end{aligned}
\tag{4.54}
$$

Grouping terms and assuming that r_0 is so much smaller than r that we may leave out terms involving powers of r_0 higher than two, leads to

$$
r_d = r \left\{ 1 - \frac{xx_0 + yy_0 + zz_0}{r^2} + \frac{x_0^2 + y_0^2 + z_0^2}{2r^2} - \frac{\left(xx_0 + yy_0 + zz_0 \right)^2}{2r^4} + \cdots \right\}.
\tag{4.55}
$$

If r is large enough relative to r_0 we only need to use the first two terms in equation (4.55) and can simplify this expression further to

$$
r_d \approx r - \frac{xx_0 + yy_0 + zz_0}{r} = r - \hat{\mathbf{u}}_{r_d} \cdot \mathbf{r}_0.
\tag{4.56}
$$

The above approximation is allowed if the third term in equation (4.55) is 'small enough'.

A widely accepted and used criterion for 'small enough' is

$$
\frac{x_0^2 + y_0^2 + z_0^2}{2r} \leq \frac{\lambda_0}{16},
\tag{4.57}
$$

where λ_0 is the free-space wavelength.[7] The criterion stated in equation (4.57) also leads to the well-known far-field criterion that states that the distance of the observation point from the origin of the antenna should satisfy

$$
r \geq \frac{2D^2}{\lambda_0},
\tag{4.58}
$$

to be in the far-field, where D is the largest dimension of the antenna.

---

[6] The tailor expansion of $\sqrt{1+x}$ is given by $\sqrt{1+x} = (1+x)^{\frac{1}{2}} = 1 + \frac{1}{2}x - \frac{1}{8}x^2 + \frac{1}{16}x^3 + \cdots$ for $(-1 < x < 1)$.
[7] Applying this criterion means that we will make a phase error of at most $k_0(\lambda_0/16) = (2\pi/\lambda_0)(\lambda_0/16) = \pi/8 = 22.5°$. This maximum phase error corresponds to the accuracy with which phase could be measured at the time this criterion was developed, which was during World War II [4].

Example

Prove that the far-field criterion of equation (4.58) follows from equation (4.57).

Let us assume an antenna with a maximum dimension D, as indicated in Figure 4.4.

The maximum value of $r_0 = |\mathbf{r}_0|$ is then equal to half that of the maximum dimension, see Figure 4.4,

$$r_0 = \sqrt{x_0^2 + y_0^2 + z_0^2} \leq \frac{D}{2}. \tag{4.59}$$

Substitution of equation (4.59) in equation (4.57) then gives

$$\frac{D^2/4}{2r} \leq \frac{\lambda_0}{16}, \tag{4.60}$$

which is equivalent with

$$r \geq \frac{2D^2}{\lambda_0}. \tag{4.61}$$

r_d in the exponent in the kernel of the integration in equation (4.52) may be replaced now by $r - \hat{\mathbf{u}}_{r_d} \cdot \mathbf{r}_0$. For the denominators in the kernel, we may even replace r_d by r since the impact on the amplitude is less severe than on the phase:

$$\frac{e^{-jk_0r_d}}{r_d} \approx \frac{e^{-jk_0r}}{r}e^{jk_0\hat{\mathbf{u}}_{r_d}\cdot\mathbf{r}_0}. \tag{4.62}$$

Further

$$\left(jk_0 + \frac{1}{r_d}\right) \approx \left(jk_0 + \frac{1}{r}\right) \approx jk_0, \tag{4.63}$$

and, see Figure 4.3,

$$\hat{\mathbf{u}}_{r_d} \approx \hat{\mathbf{u}}_r. \tag{4.64}$$

Substituting the approximations (4.62)–(4.64) in equation (4.52) finally gives us the far-field approximation for the magnetic field

$$\mathbf{H} \approx -\frac{jk_0e^{-jk_0r}}{4\pi r}\hat{\mathbf{u}}_r \times \iiint_{V_0} \mathbf{J}_e(\mathbf{r}_0)\, e^{jk_0\hat{\mathbf{u}}_r\cdot\mathbf{r}_0}dV_0. \tag{4.65}$$

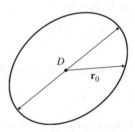

Figure 4.4 Antenna, having maximum dimension D.

4.3.2 Electric Field

The starting point for the calculation of the far electric field is equation (4.43):

$$\mathbf{E} = \frac{1}{j\omega\varepsilon_0\mu_0} \nabla \times \nabla \times \mathbf{A}_e = \frac{1}{j\omega\varepsilon_0} \nabla \times \left[\frac{1}{\mu_0}\nabla \times \mathbf{A}_e\right] = \frac{1}{j\omega\varepsilon_0}\nabla \times \mathbf{H}. \qquad (4.66)$$

Substituting equation (4.52) in equation (4.66) gives

$$\mathbf{E} = \frac{1}{j4\pi\omega\varepsilon_0} \iiint_{V_0} \nabla_r \times \left[\left(-jk_0 - \frac{1}{r_d}\right)\frac{e^{-jk_0r_d}}{r_d^2}\mathbf{r}_d \times \mathbf{J}_e\left(\mathbf{r}_0\right)\right] dV_0, \qquad (4.67)$$

where use has been made of $\hat{\mathbf{u}}_{r_d} = \mathbf{r}_d/r_d$.

To ease the solving of this equation we introduce a scalar help-variable and a vector help-variable, that we define as

$$\psi\left(\mathbf{r}, \mathbf{r}_0\right) = \left(-jk_0 - \frac{1}{r_d}\right)\frac{e^{-jk_0r_d}}{r_d^2}, \qquad (4.68)$$

$$\mathbf{a}\left(\mathbf{r}, \mathbf{r}_0\right) = \mathbf{r}_d \times \mathbf{J}_e\left(\mathbf{r}_0\right). \qquad (4.69)$$

For the sake of clarity, we will not explicitly write the $(\mathbf{r}, \mathbf{r}_0)$-dependences in ψ and $\mathbf{a}$ in the remainder of this section.

Using equation (3.52) we write for the curl in the integral of equation (4.67)

$$\nabla_r \times \psi\mathbf{a} = \nabla_r\psi \times \mathbf{a} + \psi\left[\nabla_r \times \mathbf{a}\right]. \qquad (4.70)$$

Next, we will calculate the two right-hand side components of equation (4.70). Applying the chain rule, we find

$$\nabla_r\psi = \left(-jk_0 - \frac{1}{r_d}\right)\nabla_r\left[\frac{e^{-jk_0r_d}}{r_d^2}\right] + \frac{e^{-jk_0r_d}}{r_d^2}\nabla_r\left[-jk_0 - \frac{1}{r_d}\right]$$

$$= \left(-jk_0 - \frac{1}{r_d}\right)\left[-2e^{-jk_0r_d}\frac{1}{r_d^3}\hat{\mathbf{u}}_{r_d} + \frac{1}{r_d^2}\left(-jk_0\right)e^{-jk_0r_d}\hat{\mathbf{u}}_{r_d}\right] + \frac{e^{-jk_0r_d}}{r_d^2}\left[\frac{1}{r_d^2}\hat{\mathbf{u}}_{r_d}\right]$$

$$= \frac{e^{-jk_0r_d}}{r_d^2}k_0^2\left[-1 - \frac{3}{jk_0r_d} + \frac{3}{\left(k_0r_d\right)^2}\right]\hat{\mathbf{u}}_{r_d}, \qquad (4.71)$$

and thus

$$\nabla_r\psi \times \mathbf{a} = \frac{e^{-jk_0r_d}}{r_d^2}k_0^2\left[-1 - \frac{3}{jk_0r_d} + \frac{3}{\left(k_0r_d\right)^2}\right]\hat{\mathbf{u}}_{r_d} \times \mathbf{r}_d \times \mathbf{J}_e\left(\mathbf{r}_0\right)$$

$$= \frac{e^{-jk_0r_d}}{r_d}k_0^2\left[-1 - \frac{3}{jk_0r_d} + \frac{3}{\left(k_0r_d\right)^2}\right]\hat{\mathbf{u}}_{r_d} \times \hat{\mathbf{u}}_{r_d} \times \mathbf{J}_e\left(\mathbf{r}_0\right), \qquad (4.72)$$

where, again, use has been made of $\hat{\mathbf{u}}_{r_d} = \mathbf{r}_d/r_d$.

For the second component of the right-hand-side of equation (4.70), we start with

$$\nabla_r \times \mathbf{a} = \nabla_r \times \mathbf{r}_d \times \mathbf{J}_e\left(\mathbf{r}_0\right) = \nabla_r \times \left[\left(\mathbf{r} - \mathbf{r}_0\right) \times \mathbf{J}_e\left(\mathbf{r}_0\right)\right] = -2\mathbf{J}_e\left(\mathbf{r}_0\right). \qquad (4.73)$$

Then

$$\psi \nabla_r \times \mathbf{a} = -2 \left(-jk_0 - \frac{1}{r_d} \right) \frac{e^{-jk_0 r_d}}{r_d^2} \mathbf{J}_e(\mathbf{r}_0).$$ (4.74)

Substitution of equations (4.72) and (4.74) in equation (4.70) then gives

$$\nabla_r \times \psi(\mathbf{r}, \mathbf{r}_0) \, \mathbf{a}(\mathbf{r}, \mathbf{r}_0) = \frac{e^{-jk_0 r_d}}{r_d} k_0^2 \left[-1 - \frac{3}{jk_0 r_d} + \frac{3}{(k_0 r_d)^2} \right] \hat{\mathbf{u}}_{r_d} \times \hat{\mathbf{u}}_{r_d} \times \mathbf{J}_e(\mathbf{r}_0)$$
$$+ 2 \left(jk_0 + \frac{1}{r_d} \right) \frac{e^{-jk_0 r_d}}{r_d^2} \mathbf{J}_e(\mathbf{r}_0)$$ (4.75)

We now apply the approximations for large r_d that we also applied to obtain the far magnetic field. Thus, we neglect terms r_d^{-n} for $n > 1$ and we replace r_d by $r - \hat{\mathbf{u}}_{r_d} \cdot \mathbf{r}_0$ in the exponent (phase) and replace r_d by r in the denominators (amplitude). Thus we obtain

$$\nabla_r \times \psi(\mathbf{r}, \mathbf{r}_0) \, \mathbf{a}(\mathbf{r}, \mathbf{r}_0) \approx -\frac{e^{-jk_0 r}}{r} k_0^2 e^{jk_0 \hat{\mathbf{u}}_{r_d} \cdot \mathbf{r}_0} \hat{\mathbf{u}}_{r_d} \times \hat{\mathbf{u}}_{r_d} \times \mathbf{J}_e(\mathbf{r}_0).$$ (4.76)

If we further use the approximation $\hat{\mathbf{u}}_{r_d} \approx \hat{\mathbf{u}}_r$, we finally find for the far electric field

$$\mathbf{E} \approx -\frac{k_0^2}{j\omega\varepsilon_0} \frac{e^{-jk_0 r}}{4\pi r} \hat{\mathbf{u}}_r \times \hat{\mathbf{u}}_r \times \iiint_{V_0} \mathbf{J}_e(\mathbf{r}_0) e^{jk_0 \hat{\mathbf{u}}_r \cdot \mathbf{r}_0} dV_0.$$ (4.77)

From the approximations for the far magnetic field, equation (4.65), and the far electric field, equation (4.77), we observe that

- the magnetic field is perpendicular to the propagation direction $\hat{\mathbf{u}}_r$;[8]
- the electric field is perpendicular to the propagation direction $\hat{\mathbf{u}}_r$;
- the electric field is perpendicular to the magnetic field;
- the electromagnetic waves propagate away from the source $\mathbf{J}_e(\mathbf{r}_0)$.[9]

Thus, sufficiently far away from the antenna, transverse electromagnetic (TEM) waves exist. Furthermore, the electric field may be expressed in terms of the magnetic field:

$$\mathbf{E} = \frac{-k_0^2}{j\omega\varepsilon_0 \cdot -jk_0} \hat{\mathbf{u}}_r \times \mathbf{H}$$

$$= \frac{-k_0}{\omega\varepsilon_0} \hat{\mathbf{u}}_r \times \mathbf{H}$$

$$= \frac{\omega\sqrt{\varepsilon_0 \mu_0}}{\omega\varepsilon_0} \mathbf{H} \times \hat{\mathbf{u}}_r$$

$$= Z_0 \mathbf{H} \times \hat{\mathbf{u}}_r,$$ (4.78)

where $Z_0 = \sqrt{\frac{\mu_0}{\varepsilon_0}}$ is the characteristic impedance of free space.

---

[8] The cross product of two vectors is perpendicular to both of these vectors.
[9] by virtue of the $\exp(-jk_0 r)$ term.

Equations (4.65) and (4.77) also show that the far magnetic and electric fields at an observation point consist of the added contributions of all source points within the source volume V_0, corrected for their relative positions within V_0 with a factor $\exp\left(jk_0\hat{\mathbf{u}}_r \cdot \mathbf{r}_0\right)$. This last term creates phase-differences between the various source-point contributions. Therefore, in the far-field we may find positions where constructive interference of the source-point contributions occurs and positions where destructive interference of the source-point contributions occurs. Therefore, the three-dimensional radiation pattern will not be uniform (although two-dimensional pattern cuts may be, as we have seen in Chapter 2).

In Chapter 2 we stated the far electric field in equations (2.2)–(2.4) without any proof. Equation (4.77) now shows that the amplitude and phase behavior, $\exp\left(-jk_0r\right)/r$, is correct and that, since $\mathbf{E}$ is perpendicular to $\hat{\mathbf{u}}_r$, it is of the form $\mathbf{E} = E_\vartheta\hat{\mathbf{u}}_\vartheta + E_\varphi\hat{\mathbf{u}}_\varphi$.

The vector of Poynting (time-averaged power density) is then found to be

$$\mathbf{S} = \frac{1}{2Z_0}\,|\mathbf{E}|^2\,\hat{\mathbf{u}}_r. \tag{4.79}$$

4.4 Reciprocity

So far we have implicitly assumed that antennas are being used for transmission of electromagnetic waves, using a current density as a source. Through the *Lorentz reciprocity theorem* we will show that antennas, when used for receiving electromagnetic waves, exhibit the same characteristics as when being used for transmitting electromagnetic waves. The receiving antenna pattern (a sensitivity pattern) is identical to the transmitting antenna pattern (a radiation pattern).

The Lorentz reciprocity theorem states that the relationship between a time-varying current density and the resulting electric field is unchanged if one interchanges the points where the current density is placed and where the field is measured.

4.4.1 Lorentz Reciprocity Theorem

So, let's consider two volumes V_A and V_B with source current densities $\mathbf{J}_A$ and $\mathbf{J}_B$, respectively, see Figure 4.5.

Source $\mathbf{J}_A$ creates fields $\mathbf{E}_A$ and $\mathbf{H}_A$ everywhere in space. The fields are interrelated through

$$\nabla \times \mathbf{H}_A = j\omega\varepsilon_0\mathbf{E}_A + \mathbf{J}_A, \tag{4.80}$$

and

$$\nabla \times \mathbf{E}_A = -j\omega\mu_0\mathbf{H}_A. \tag{4.81}$$

Similarly, source $\mathbf{J}_B$ creates fields $\mathbf{E}_B$ and $\mathbf{H}_B$ everywhere in space. These fields are interrelated through

$$\nabla \times \mathbf{H}_B = j\omega\varepsilon_0\mathbf{E}_B + \mathbf{J}_B, \tag{4.82}$$

$$\nabla \times \mathbf{E}_B = -j\omega\mu_0\mathbf{H}_B. \tag{4.83}$$

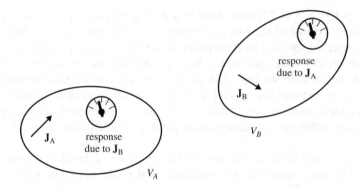

Figure 4.5   Source $\mathbf{J}_A$ in volume V_A creates fields $\mathbf{E}_A$ and $\mathbf{H}_A$ everywhere in space and source $\mathbf{J}_B$ in volume V_B creates fields $\mathbf{E}_B$ and $\mathbf{H}_B$ everywhere in space.

We now form the dot-product of equation (4.80) with $\mathbf{E}_B$, the dot-product of equation (4.83) with $\mathbf{H}_A$ and subtract the results, which leads to

$$-\nabla \cdot (\mathbf{E}_B \times \mathbf{H}_A) = j\omega\varepsilon_0\mathbf{E}_A \cdot \mathbf{E}_B + j\omega\mu_0\mathbf{H}_A \cdot \mathbf{H}_B + \mathbf{E}_B \cdot \mathbf{J}_A, \qquad (4.84)$$

where we have made use of the vector identity

$$\nabla \cdot (\mathbf{a} \times \mathbf{b}) = \mathbf{b} \cdot \nabla \times \mathbf{a} - \mathbf{a} \cdot \nabla \times \mathbf{b}. \qquad (4.85)$$

Now, we interchange the subscripts A and B in equation (4.84) and subtract the original equation from this new equation. This results in

$$-\nabla \cdot (\mathbf{E}_A \times \mathbf{H}_B - \mathbf{E}_B \times \mathbf{H}_A) = \mathbf{E}_A \cdot \mathbf{J}_B - \mathbf{E}_B \cdot \mathbf{J}_A. \qquad (4.86)$$

Outside the source volumes V_A and V_B, $\mathbf{J}_A = \mathbf{J}_B = 0$, see Figure 4.5, and equation (4.86) reduces to

$$-\nabla \cdot (\mathbf{E}_A \times \mathbf{H}_B - \mathbf{E}_B \times \mathbf{H}_A) = 0. \qquad (4.87)$$

This equation is known as the *Lorentz reciprocity theorem*.

The integral form of the Lorentz reciprocity theorem for a source-free volume is found by integrating equation (4.87) over the source-free volume and then applying the divergence theorem.[10] This results in

$$\iint_S (\mathbf{E}_A \times \mathbf{H}_B - \mathbf{E}_B \times \mathbf{H}_A) \cdot d\mathbf{S} = 0. \qquad (4.89)$$

For a volume containing sources, integration of equation (4.86) over that volume results in

$$-\iint_S (\mathbf{E}_A \times \mathbf{H}_B - \mathbf{E}_B \times \mathbf{H}_A) \cdot d\mathbf{S} = \iiint_V (\mathbf{E}_A \cdot \mathbf{J}_B - \mathbf{E}_B \cdot \mathbf{J}_A) \, dV. \qquad (4.90)$$

---

[10] Divergence theorem:

$$\iiint_V \nabla\psi \, dV = \iint_S \psi \, d\mathbf{S}. \qquad (4.88)$$

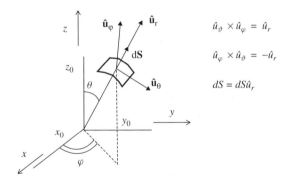

Figure 4.6 Spherical coordinates and unit vectors.

Next, we assume that our sources are finite in extent, as antennas are. We have seen that far away from the sources, the electric and magnetic fields are locally TEM. This means that of the three possible spherical coordinates, only ϑ- and φ-components actually exist, see Figure 4.6.

The field components at large distances from the source thus behave – see Figure 4.6 and equation (4.78) – as

$$E_\vartheta = Z_0 H_\varphi, \tag{4.91}$$

$$E_\varphi = -Z_0 H_\vartheta. \tag{4.92}$$

If we substitute this in the left-hand side of equation (4.90) we obtain

$$-Z_0 \iint_S \left(H_{A_\vartheta} H_{B_\vartheta} + H_{A_\varphi} H_{B_\varphi} - H_{B_\vartheta} H_{A_\vartheta} - H_{B_\varphi} H_{A_\varphi} \right) dS = 0, \tag{4.93}$$

reducing equation (4.90) to

$$\iiint_{V_B} \mathbf{E}_A \cdot \mathbf{J}_B dV = \iiint_{V_A} \mathbf{E}_B \cdot \mathbf{J}_A dV. \tag{4.94}$$

This relation between sources and fields will be used to prove antenna reciprocity.

4.4.2 Antenna Reciprocity

Consider two perfect electrically conducting antennas, A and B. An electric field will only be present between the clamps of each antenna. If we also pose the currents to be constant through the clamps, equation (4.94) will become

$$V_A I_A = V_B I_B, \tag{4.95}$$

where V_A is the open circuit voltage across the clamps of antenna A, due to an electric field $\mathbf{E}_B$. V_B is the open circuit voltage across the clamps of antenna B, due to an electric field $\mathbf{E}_A$.[11]

---

[11] Remember that $\int \mathbf{E} \cdot d\mathbf{l} = -V$.

We may write equation (4.95) as

$$\frac{V_A}{I_B} = \frac{V_B}{I_A}.$$

(4.96)

In general, the voltage at the clamps of one antenna, due to another antenna that is excited, is influenced by a number of factors. These factors include both antennas, the medium, obstacles, and so on. In circuit parameters we may write this as

$$V_A = Z_{AA}I_A + Z_{AB}I_B,$$

(4.97)

$$V_B = Z_{BA}I_A + Z_{BB}I_B,$$

(4.98)

where V_A, V_B, I_A and I_B are the clamp voltages and currents of antennas A and B.

When antenna A is excited with a current I_A, the open circuit voltage of antenna B is $V_B|_{I_B=0}$. The transfer impedance Z_{BA} then follows from

$$Z_{BA} = \frac{V_B}{I_A}\bigg|_{I_B=0}.$$

(4.99)

When antenna B is excited with a current I_B, the open circuit voltage of antenna A is $V_A|_{I_A=0}$. The transfer impedance Z_{AB} then follows from

$$Z_{AB} = \frac{V_A}{I_B}\bigg|_{I_A=0}.$$

(4.100)

If we compare equations (4.96), (4.99) and (4.100), we see that as a result of the reciprocity theorem

$$Z_{BA} = Z_{AB} = Z_m,$$

(4.101)

where Z_m is the transfer impedance between the two antennas.

We assume now that antenna A is excited with current I_A and that the voltage V_B is measured across the clamps of antenna B. We also assume that both antennas are spaced

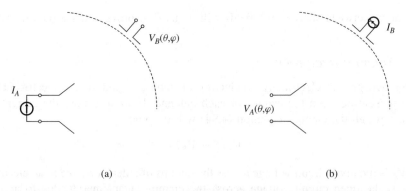

(a)   (b)

Figure 4.7   The transmit and receive patterns of an antenna are identical since $Z_{AB}(\vartheta, \varphi) = Z_{BA}(\vartheta, \varphi) = Z_m(\vartheta, \varphi)$. (a) The transmission pattern of antenna A is $Z_{BA}(\vartheta, \varphi) = V_B(\vartheta, \varphi)/I_A$. (b) The receive pattern of antenna A is $Z_{AB}(\vartheta, \varphi) = V_A(\vartheta, \varphi)/I_B$.

apart over such a distance that they are positioned in each other's far-field regions. The transfer impedance Z_{BA} is now the far-field radiation pattern of antenna A, if antenna B is moved around antenna A over the surface of a sphere with a constant radius, see Figure 4.7(a).

If now antenna B is excited and antenna A is used for reception, the transfer impedance Z_{AB} is the receive pattern, if antenna B is moved over the same sphere.

Since the transfer impedances Z_{AB} and Z_{BA} are equal, a consequence of the reciprocity theorem is that the transmit and receive patterns of an antenna are identical.

4.5 Problems

4.1 Derive the so-called continuity equation $\nabla \cdot \mathbf{J}_e = -\frac{\partial \rho_e}{\partial t}$ from the Maxwell equations. Start by taking the divergence of equation (4.2).

4.2 Prove that the solutions for $\mathbf{A}_e$, equation (4.36), and ϕ_e, equation (4.37), still satisfy the Lorentz gauge $\phi_e = -\frac{1}{j\omega\varepsilon_0\mu_0}$.

4.3 If, in approximating the far-field, we would not accept phase errors exceeding 10 degrees, what would be the new rule of thumb for the far-field (instead of $r \geq \frac{2D^2}{\lambda_0}$)?

4.4 With $\mathbf{S}(\mathbf{r}) = \frac{1}{2}\Re\{\mathbf{E}(\mathbf{r}) \times \mathbf{H}^*(\mathbf{r})\}$, prove that $\mathbf{S}(\mathbf{r}) = \frac{1}{2Z_0}|\mathbf{E}(\mathbf{r})|^2\,\hat{\mathbf{u}}_r$.

4.5 Explain why, for the explanation of antenna reciprocity – see Section 4.4.2 and Figure 4.7 – it is necessary that the space in between the two antennas is free of (fixed) obstacles.

References

1. Simon Ramo, John R. Whinnery and Theodore van Duzer, *Fields and Waves in Communication Electronics*, *third edition*, John Wiley & Sons, New York, 1994.
2. John D. Kraus and Daniel Fleisch, *Electromagnetics*, *fifth edition*, McGraw-Hill, 1999.
3. Milton Abramowitz and Irene A. Stegun, *Handbook of Mathematical Functions*, Dover Publications, New York, 1972.
4. C.G. Montgomery, *Technique of Microwave Measurements*, Volume 11 of MIT Radiation Laboratory Series, McGraw-Hill, New York, 1947.

5

Dipole Antennas

In the previous chapter, we started with an infinitesimal, z-directed, constant current[1] to derive the vector potential. From there we generalized the vector potential by allowing an arbitrary orientation and the current to be distributed over a source volume. For deriving the dipole radiation pattern, we will start again with the infinitesimal, z-directed, constant current. From there, we will lengthen the dipole, include a feeding point and from an assessment of the current distribution calculate the radiation pattern. As well as the radiation pattern, from a system point of view, we also want to know the input impedance. Since the derivation of the input impedance is beyond the scope of this book, we will summarize a few approximate equations for this parameter.

5.1 Elementary Dipole

The starting point of the dipole antenna discussion will be the elementary or Hertz dipole. This elementary dipole is (here) a z-directed current with constant amplitude I_0 over infinitesimal length l, see Figure 5.1(a), that we also used to start the discussion of the vector potential and radiation fields, see Figure 4.1.

Before we proceed, we must make some remarks concerning Figure 5.1(a).

- First of all, we note that this elementary dipole antenna cannot exist stand-alone in reality since the current needs somewhere to come from and somewhere to go to. We may think of this elementary dipole as a small length of current-carrying conductor that is one of multiple segments into which we must divide a real (wire) antenna, in order to calculate its properties. This imaginary elementary dipole antenna is important, however, because we can easily calculate the far electric and magnetic fields.
- The name of this elementary radiator, *dipole*, suggests that we are dealing with a doublet of equal-amplitude, positive and negative charges. To demonstrate that this is indeed so, we start by taking the divergence of the Maxwell equation (4.12). Using the vector identity (3.49) then leads to

$$\nabla \cdot \nabla \times \mathbf{H}(\mathbf{r}) = 0 = j\omega\varepsilon_0 \nabla \cdot \mathbf{E}(\mathbf{r}) + \nabla \cdot \mathbf{J}_e(\mathbf{r}). \tag{5.1}$$

[1] By constant we mean that the amplitude of the current is constant over the length of the current. The current itself is oscillatory, having an $\exp(j\omega t)$ time dependence.

Antenna Theory and Applications, First Edition. Hubregt J. Visser.
© 2012 John Wiley & Sons, Ltd. Published 2012 by John Wiley & Sons, Ltd.

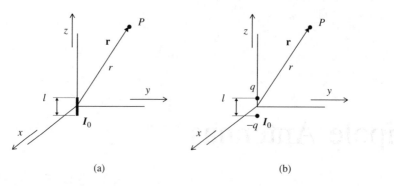

(a)   (b)

Figure 5.1 Elementary or Hertz dipole. (a) z-directed current with constant amplitude I_0 over infinitesimal length l. (b) Two equal, but opposite charges, displaced over a distance l along the z-axis of a rectangular coordinate system.

Next, we substitute the Maxwell equation (4.14) in equation (5.1) and obtain

$$\rho_e(\mathbf{r}) = -\frac{1}{j\omega} \nabla \cdot \mathbf{J}_e(\mathbf{r}).$$ (5.2)

The current density is given by, see Figure 5.1(a),

$$\mathbf{J}_e(\mathbf{r}) = I_0 l \delta(x) \delta(y) \delta(z) \hat{\mathbf{u}}_z.$$ (5.3)

The derivative of the Dirac delta follows from the definition of the derivative,

$$\frac{d\delta(z)}{dz} = \lim_{l \to 0} \frac{\delta\left(z + \frac{l}{2}\right) - \delta\left(z - \frac{l}{2}\right)}{l},$$ (5.4)

so that equation (5.4) substituted in equation (5.2) gives

$$\rho_e(\mathbf{r}) = -\lim_{l \to 0} \frac{I_0 \delta(x) \delta(y)}{j\omega} \left[\delta\left(z + \frac{l}{2}\right) - \delta\left(z - \frac{l}{2}\right) \right].$$ (5.5)

We see that the charge density is indeed that of a *di*pole, the charges being located at $(x, y, z) = (0, 0, l/2)$ and $(x, y, z) = (0, 0, -l/2)$.

5.1.1 Radiation

As we saw in the previous chapter, the far magnetic and electric fields may be approximated[2] by equations (4.65) and (4.78):

$$\mathbf{H} = \frac{-jk_0 e^{-jk_0 r}}{4\pi r} \hat{\mathbf{u}}_r \times \iiint_{V_0} \mathbf{J}_e(\mathbf{r}_0) e^{jk_0 \hat{\mathbf{u}}_r \cdot \mathbf{r}_0} dV_0,$$ (5.6)

$$\mathbf{E} = Z_0 \mathbf{H} \times \hat{\mathbf{u}}_r,$$ (5.7)

where $Z_0 = \sqrt{\frac{\mu_0}{\varepsilon_0}}$ is the characteristic impedance of free space.

---

[2] Since in calculating the radiation patterns of antennas we will always be in the far-field, we replace the $\approx$-sign in the far-field approximations with an equality sign.

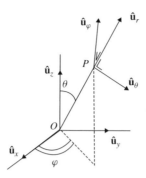

Figure 5.2 Rectangular coordinate unit vectors, relative to the origin O and spherical coordinate unit vectors, relative to the observation point P.

Upon substituting equation (5.3) in equation (5.6) we need to calculate a cross product of $\hat{\mathbf{u}}_r$ and $\hat{\mathbf{u}}_z$. To perform this cross product calculation, we will first transform the rectangular coordinate unit vector into a combination of spherical coordinate unit vectors using [1], see also Figure 5.2,

$$\hat{\mathbf{u}}_x = \sin\vartheta\cos\varphi\hat{\mathbf{u}}_r + \cos\vartheta\cos\varphi\hat{\mathbf{u}}_\vartheta - \sin\varphi\hat{\mathbf{u}}_\varphi, \tag{5.8}$$

$$\hat{\mathbf{u}}_y = \sin\vartheta\sin\varphi\hat{\mathbf{u}}_r + \cos\vartheta\sin\varphi\hat{\mathbf{u}}_\vartheta + \cos\varphi\hat{\mathbf{u}}_\varphi, \tag{5.9}$$

$$\hat{\mathbf{u}}_z = \cos\vartheta\hat{\mathbf{u}}_r - \sin\vartheta\hat{\mathbf{u}}_\vartheta. \tag{5.10}$$

With equation (5.10) we find that $\hat{\mathbf{u}}_r \times \hat{\mathbf{u}}_z = \cos\vartheta\hat{\mathbf{u}}_r \times \hat{\mathbf{u}}_r - \sin\vartheta\hat{\mathbf{u}}_r \times \hat{\mathbf{u}}_\vartheta = -\sin\vartheta\hat{\mathbf{u}}_r \times \hat{\mathbf{u}}_\vartheta$. With the aid of Figure 5.2 we find that $\hat{\mathbf{u}}_r \times \hat{\mathbf{u}}_\vartheta = \hat{\mathbf{u}}_\varphi$, so that

$$\mathbf{H} = \frac{-jk_0 e^{-jk_0 r}}{4\pi r} \iiint_{V_0} -I_0 l \delta\,(x_0)\,\delta\,(y_0)\,\delta\,(z_0) \sin\vartheta\hat{\mathbf{u}}_\varphi e^{jk_0\hat{\mathbf{u}}_r\cdot\mathbf{r}_0} dV_0$$

$$= \frac{jk_0 e^{-jk_0 r}}{4\pi r} I_0 l \sin\vartheta\hat{\mathbf{u}}_\varphi, \tag{5.11}$$

and, since $\hat{\mathbf{u}}_\varphi \times \hat{\mathbf{u}}_r = \hat{\mathbf{u}}_\vartheta$, see Figure 5.2,

$$\mathbf{E} = Z_0\mathbf{H} \times \hat{\mathbf{u}}_r = \frac{jk_0 Z_0 I_0 l e^{-jk_0 r}}{4\pi r} \sin\vartheta\hat{\mathbf{u}}_\vartheta. \tag{5.12}$$

Now we can calculate the radiation pattern.

To start with, we find the time-average power density per unit of surface area $\left(r^2\sin\vartheta\,d\vartheta\,d\varphi = r^2 d\Omega\right)$ by substituting the expression for the electric field in equation (2.8):

$$\mathbf{S(r)} = \mathbf{S}(\vartheta,\varphi) = \frac{1}{2Z_0}|\mathbf{E(r)}|^2\,\hat{\mathbf{u}}_r = \frac{1}{2}\frac{k_0^2 Z_0\,(I_0 l)^2}{(4\pi r)^2}\sin^2\vartheta\hat{\mathbf{u}}_r. \tag{5.13}$$

The radiated power per unit of solid angle $(d\Omega)$ is then

$$P\,(\vartheta,\varphi) = \left|r^2\mathbf{S}\,(\vartheta,\varphi)\right| = \frac{1}{2}\frac{k_0^2 Z_0\,(I_0 l)^2}{(4\pi)^2}\sin^2\vartheta. \tag{5.14}$$

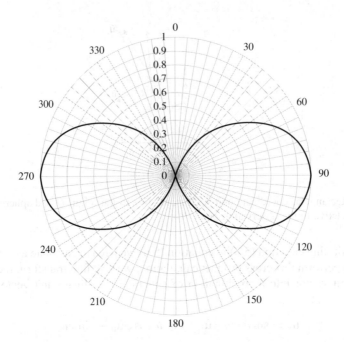

Figure 5.3 Polar radiation pattern cut in a plane $\varphi = $ constant on a linear amplitude scale.

The maximum radiated power is found for $\vartheta = \frac{\pi}{2}$, so that the normalized radiation pattern is given by

$$F(\vartheta, \varphi) = F(\vartheta) = \frac{P(\vartheta)}{P_{max}} = \frac{P(\vartheta)}{P\left(\frac{\pi}{2}\right)} = \sin^2 \vartheta. \qquad (5.15)$$

The radiation pattern does not show a φ-dependence; the pattern is rotationally symmetric around the z-axis of the rectangular coordinate system. A polar plot of the radiation pattern in an arbitrary plane φ, on a linear scale, is shown in Figure 5.3.

As expected, we observe that the radiation is at maximum in the directions perpendicular to the dipole and that no radiation occurs in the directions along the dipole. For completeness, in Figure 5.4(a) we show the $\varphi = $ constant radiation pattern cut in a rectangular plot, having a linear amplitude scale, and in Figure 5.4(b) we show the pattern cut in a rectangular plot, having a logarithmic ($10 \log [F(\vartheta)]$) scale.

With the radiation pattern now shown, we will proceed with quantifying the radiation pattern, that is calculate the directivity. We defined the directivity function in Chapter 2, equation (2.9), as

$$D(\vartheta, \varphi) = \frac{P(\vartheta, \varphi)}{P_t/4\pi}, \qquad (5.16)$$

where P_t is the total transmitted power. To calculate the total radiated power we integrate the radiated power per unit of solid angle, $P(\vartheta)$, over all possible angles ϑ and φ. With

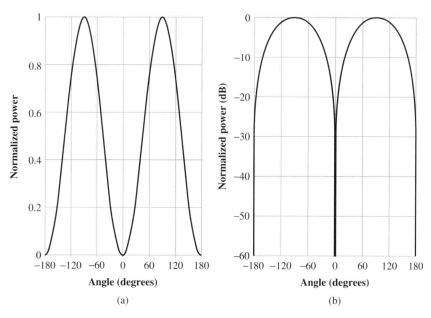

Figure 5.4   Rectangular radiation pattern cut in a plane φ = constant. (a) Linear amplitude scale. (b) Logarithmic amplitude scale.

the aid of equation (5.14), we find

$$P_t = \int_{\varphi=0}^{2\pi} \int_{\vartheta=0}^{\pi} P(\vartheta) \sin\vartheta \, d\vartheta \, d\varphi$$

$$= \frac{k_0^2 Z_0 (I_0 l)^2}{32\pi^2} \int_{\varphi=0}^{2\pi} \int_{\vartheta=0}^{\pi} \sin^3(\vartheta) d\vartheta \, d\varphi$$

$$= \frac{k_0^2 Z_0 (I_0 l)^2}{16\pi} \int_{\vartheta=0}^{\pi} \sin^3(\vartheta) d\vartheta =$$

$$= \frac{k_0^2 Z_0 (I_0 l)^2}{12\pi}. \tag{5.17}$$

In the above, use has been made of $\int_{\vartheta=0}^{\pi} \sin^3(\vartheta) d\vartheta = \frac{4}{3}$.

Substitution of equations (5.14) and (5.17) in equation (5.16) gives the directivity function as

$$D(\vartheta) = \frac{3}{2} \sin^2(\vartheta), \tag{5.18}$$

and the directivity, being defined as the maximum of the directivity function, is therefore

$$D = [D(\vartheta)]_{max} = \frac{3}{2}, \tag{5.19}$$

meaning that an elementary dipole radiates – in the direction of maximum radiation – one and a half times the power that a (hypothetical) uniform radiator would, for the same total radiated power.[3]

5.1.2 Input Impedance

To avoid needlessly complicating the analysis, we assume that the elementary dipole only has a real input impedance and no losses so that the input impedance equals the radiation resistance. From equation (2.17) we know that the total radiated power may be expressed then as

$$P_t = \frac{\frac{1}{2}|V_A|^2}{R_A}, \tag{5.20}$$

where V_A is the voltage over the antenna clamps and R_A is the radiation resistance. Since $V_A = I_0 R_A$, we find for the radiation resistance

$$R_A = \frac{P_t}{\frac{1}{2}|I_0|^2}. \tag{5.21}$$

Substitution of equation (5.17) in equation (5.21) and using $k_0 = \frac{2\pi}{\lambda_0}$, λ_0 being the wavelength in free space, and $Z_0 = 120\pi\,\Omega$ gives for the radiation resistance

$$R_A = 80\pi^2 \left(\frac{l}{\lambda_0}\right)^2. \tag{5.22}$$

Note that this equation is valid for an elementary dipole, thus for $l \downarrow 0$, only.

Example

Calculate the input impedance of a small dipole antenna operating at 3000 kHz and having a length of 1 m. Calculate the reflection coefficient when an antenna having this input impedance is connected to a 50 Ω transmission line.

Assuming no losses and all impedances being real, the input impedance of an elementary dipole antenna will equal the radiation resistance. For the frequency and length stated, the dipole antenna length equals $l = 0.01\lambda_0$ which is indeed a small or elementary dipole antenna. Substituting this length in equation (5.22) results in

$$R_A = Z_{in} = 80\pi^2 (0.01)^2 = 0.079\,\Omega. \tag{5.23}$$

The reflection coefficient is found upon substituting this input impedance value in equation (2.20)

$$\Gamma = \left|\frac{R_A - Z_0}{R_A + Z_0}\right| = \left|\frac{R_A - 50}{R_A + 50}\right| = \left|\frac{0.079 - 50}{0.079 + 50}\right| = 0.9968. \tag{5.24}$$

The example shows that the input impedance of an elementary dipole antenna is very small, which makes it very difficult to impedance match it to standard (50 Ω) equipment.

---

[3] Since, in comparing with a uniform radiator, the same total power is radiated, the 'gain' of one and a half in certain directions must be compensated for in other directions. This explains the 'nulls' in directions along the dipole axis.

In real life we also have to deal with (ohmic) losses and, due to the low radiation resistance, the losses (unwanted) will easily dominate the radiation resistance (wanted loss) so that a very small radiator in general will have a very low radiation efficiency.[4] Equation (5.22) shows that we can increase the efficiency, by increasing the radiation resistance as a result of increasing the dipole length. However, equation (5.22) is only valid for small lengths, so we have to derive new equations for a non-infinitesimal dipole antenna.

5.2 Non-Infinitesimal Dipole Antenna

We now consider a thin wire antenna of non-infinitesimal half-length l, excited in the middle, see Figure 5.5.

We assume that $a \ll \lambda_0$ and $d \ll \lambda_0$, and therefore the influences from the radius and the feed gap may be neglected. The current density needs to be zero at $z = l$ and at $z = -l$. Also, since the current along the wires of the antenna should be essentially the same as that on an open-circuited two-wire transmission line (we may think of the wire antenna in Figure 5.5 as being constructed by bending each of the ends of a two-wire transmission line 90 degrees outward), the current density will be sinusoidal and may be written as

$$\mathbf{J}_e = I_0 \delta(x)\delta(y) \sin\left[k_0\left(l - |z|\right)\right] \hat{\mathbf{u}}_z \text{ for } -l \le z \le l. \tag{5.25}$$

5.2.1 Radiation

For finding the normalized radiation pattern and calculating the directivity, we first need to calculate the far electric field. Therefore we use equation (4.77)

$$\mathbf{E} = \frac{-k_0^2}{j\omega\varepsilon_0} \frac{e^{-jk_0 r}}{4\pi r} \hat{\mathbf{u}}_r \times \hat{\mathbf{u}}_r \times \iiint_{V_0} \mathbf{J}_e(\mathbf{r}_0) e^{jk_0 \hat{\mathbf{u}}_r \cdot \mathbf{r}_0} dV_0. \tag{5.26}$$

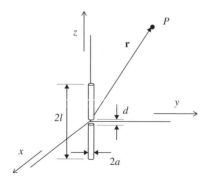

Figure 5.5 z-directed, thin-wire antenna of half-length l.

---

[4] The radiation efficiency η is given by

$$\eta = \frac{R_A}{R_A + R_L}, \tag{5.27}$$

where R_L is the loss resistance.

After substituting equation (5.25) in equation (5.26), we need to calculate $\hat{\mathbf{u}}_r \times \hat{\mathbf{u}}_r \times \hat{\mathbf{u}}_z$. To be able to perform this calculation we will first transform the rectangular coordinate system unit vector $\hat{\mathbf{u}}_z$ into spherical coordinate system unit vectors, using equation (5.10). Then, also using Figure 5.2,

$$\hat{\mathbf{u}}_r \times \hat{\mathbf{u}}_r \times \hat{\mathbf{u}}_z = \hat{\mathbf{u}}_r \times \hat{\mathbf{u}}_r \times \left[\cos \vartheta \hat{\mathbf{u}}_r - \sin \vartheta \hat{\mathbf{u}}_\vartheta\right] = -\sin \vartheta \hat{\mathbf{u}}_r \times \hat{\mathbf{u}}_\varphi = \sin \vartheta \hat{\mathbf{u}}_\vartheta. \quad (5.28)$$

Also by using equation (5.10) and Figure 5.2 the dot product in the exponent in the integral is found to be

$$\hat{\mathbf{u}}_r \cdot \mathbf{r}_0 = \hat{\mathbf{u}}_r \cdot z_0 \hat{\mathbf{u}}_z = \hat{\mathbf{u}}_r \cdot z_0 \left[\cos \vartheta \hat{\mathbf{u}}_r - \sin \vartheta \hat{\mathbf{u}}_\vartheta\right] = z_0 \cos \vartheta, \quad (5.29)$$

so that, finally, equation (5.26) becomes

$$\mathbf{E} = \frac{-k_0^2 I_0}{j\omega\varepsilon_0} \frac{e^{-jk_0 r}}{4\pi r} \sin \vartheta \hat{\mathbf{u}}_\vartheta \int_{-l}^{l} \sin \left[k_0 \left(l - |z_0|\right)\right] e^{jk_0 z_0 \cos \vartheta} dz_0. \quad (5.30)$$

We see that the far electric field of the z-directed, thin wire antenna only has a ϑ-component, just as we found for the infinitesimal, z-directed dipole antenna, equation (5.12).

We proceed by 'processing' the integral in equation (5.30). We start by expanding the $|z_0|$-term:

$$I = \int_{-l}^{l} \sin \left[k_0 \left(l - |z_0|\right)\right] e^{jk_0 z_0 \cos \vartheta} dz_0$$

$$= \int_{-l}^{0} \sin \left[k_0 \left(l + z_0\right)\right] e^{jk_0 z_0 \cos \vartheta} dz_0 + \int_{0}^{l} \sin \left[k_0 \left(l - z_0\right)\right] e^{jk_0 z_0 \cos \vartheta} dz_0. \quad (5.31)$$

Then, in the first integral on the right-hand-side of equation (5.31) we change z_0 into $-z_0$ and, at the same time, we interchange the integration boundaries, so that we obtain

$$I = \int_{0}^{l} \sin \left[k_0 \left(l - z_0\right)\right] e^{-jk_0 z_0 \cos \vartheta} dz_0 + \int_{0}^{l} \sin \left[k_0 \left(l - z_0\right)\right] e^{jk_0 z_0 \cos \vartheta} dz_0$$

$$= 2 \int_{0}^{l} \sin \left[k_0 \left(l - z_0\right)\right] \frac{e^{jk_0 z_0 \cos \vartheta} + e^{-jk_0 z_0 \cos \vartheta}}{2} dz_0$$

$$= 2 \int_{0}^{l} \sin \left[k_0 \left(l - z_0\right)\right] \cos \left[k_0 z_0 \cos \vartheta\right] dz_0. \quad (5.32)$$

Next, using $2 \sin A \cos B = \sin[A + B] + \sin[A - B]$, we obtain

$$I = \int_{0}^{l} \sin \left[k_0 l + z_0 k_0(-1 + \cos \vartheta)\right] dz_0 + \int_{0}^{l} \sin \left[k_0 l - z_0 k_0(1 + \cos \vartheta)\right] dz_0$$

$$= \frac{2 \left\{\cos \left[k_0 l \cos \vartheta\right] - \cos \left[k_0 l\right]\right\}}{k_0 \sin^2 \vartheta}. \quad (5.33)$$

Upon substitution of equation (5.33) for the integral in equation (5.30), we find for the far electric field

$$\mathbf{E} = jZ_0I_0\frac{e^{-jk_0r}}{2\pi r}\left[\frac{\cos(k_0l\cos\vartheta) - \cos(k_0l)}{\sin\vartheta}\right]\hat{\mathbf{u}}_\vartheta,\tag{5.34}$$

where we have used

$$\frac{1}{\omega\varepsilon_0} = \frac{1}{\omega\sqrt{\varepsilon_0\mu_0}\sqrt{\frac{\varepsilon_0}{\mu_0}}} = \frac{\sqrt{\frac{\mu_0}{\varepsilon_0}}}{k_0} = \frac{Z_0}{k_0}.\tag{5.35}$$

Both for calculating the radiation pattern and for calculating the gain, we need to know the radiated power per unit of solid angle. This quantity is found, using equation (2.8) and equation (5.34) as

$$P(\vartheta) = \left|r^2S(\vartheta)\right| = \frac{r^2|\mathbf{E}|^2}{2Z_0} = \frac{Z_0I_0^2}{8\pi^2}\left[\frac{\cos(k_0l\cos\vartheta) - \cos(k_0l)}{\sin\vartheta}\right]^2.\tag{5.36}$$

The normalized radiation pattern is obtained by dividing by the maximum of $P(\vartheta)$. We will show the normalized radiation patterns for a number of antenna lengths $2l$, starting with the important (i.e., much-used) half-wavelength antenna ($2l = \lambda_0/2$).

5.2.1.1 Half-Wavelength Antenna

For $2l = \lambda_0/2$, remembering that $k_0 = 2\pi/\lambda_0$, equation (5.36) reduces to

$$P(\vartheta) = \frac{Z_0I_0^2}{8\pi^2}\left[\frac{\cos\left(\frac{\pi}{2}\cos\vartheta\right)}{\sin\vartheta}\right]^2.\tag{5.37}$$

The maximum is found for $\vartheta = \frac{\pi}{2}$ and equals $\frac{Z_0I_0^2}{8\pi^2}$ so that we find for the normalized (power) radiation pattern

$$F(\vartheta) = \left[\frac{\cos\left(\frac{\pi}{2}\cos\vartheta\right)}{\sin\vartheta}\right]^2.\tag{5.38}$$

A polar, linear plot of the radiation pattern in a plane $\varphi = $ constant is shown in Figure 5.6. In the same figure we also show, for comparison the radiation pattern cut for an elementary dipole antenna.

The figure clearly shows that the radiation pattern of a half-wave dipole antenna is very similar to that of an elementary dipole antenna. The lobes are slightly narrower than those of an elementary dipole antenna, so the directivity should be slightly larger than the directivity of an elementary dipole ($D = 1.5$).

For calculating the directivity function, we need to know the total radiated power. Therefore we need to integrate equation (5.37) over all angles φ and ϑ.

The total radiated power is calculated as

$$P_t = \int_{\varphi=0}^{2\pi}\int_{\vartheta=0}^{\pi}P(\vartheta)d\Omega = 2\pi\int_0^\pi P(\vartheta)\sin\vartheta d\vartheta.\tag{5.39}$$

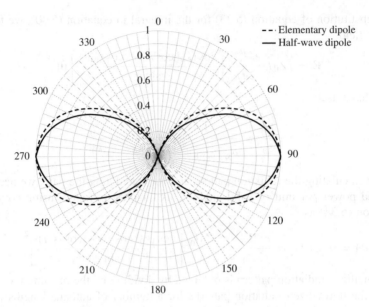

Figure 5.6 Polar radiation pattern cuts in a plane $\varphi = $ constant on a linear amplitude scale. Solid line: half-wave dipole antenna. Dashed line: elementary dipole antenna.

Substitution of equation (5.37) in equation (5.39) gives

$$P_t = \frac{Z_0 I_0^2}{4\pi} \int_0^\pi \left[\frac{\cos\left(\frac{\pi}{2}\cos\vartheta\right)}{\sin\vartheta} \right]^2 \sin\vartheta \, d\vartheta. \qquad (5.40)$$

We will solve this equation by casting it into a form for which a solution is known or tabulated.

The first step is substituting $\cos\vartheta = u$. In doing so, $\cos^2\left(\frac{\pi}{2}\cos\vartheta\right) = \cos^2\left(\frac{\pi u}{2}\right)$, $\sin^2\vartheta = 1 - \cos^2\vartheta = 1 - u^2$ and $\sin\vartheta \, d\vartheta = -du$. Thus, equation (5.40) transforms into

$$P_t = -\frac{Z_0 I_0^2}{4\pi} \int_{u=1}^{-1} \frac{\cos^2\left(\frac{\pi u}{2}\right)}{1 - u^2} du = \frac{Z_0 I_0^2}{4\pi} \int_{-1}^{1} \frac{\cos^2\left(\frac{\pi u}{2}\right)}{1 - u^2} du. \qquad (5.41)$$

Next, we substitute $1/\left(1 - u^2\right) = 1/(1-u)(1+u) = \frac{1}{2}/(1-u) + \frac{1}{2}/(1+u)$ to obtain

$$P_t = \frac{Z_0 I_0^2}{8\pi} \int_{-1}^{1} \frac{\cos^2\left(\frac{\pi u}{2}\right)}{1 - u} du + \frac{Z_0 I_0^2}{8\pi} \int_{-1}^{1} \frac{\cos^2\left(\frac{\pi u}{2}\right)}{1 + u} du. \qquad (5.42)$$

Then, we substitute in the first integral in equation (5.42) $1 - u = s/\pi$, so that $u = (\pi - s)/\pi$ and $du = d(-s)/\pi$. In the second integral, we substitute $1 + u = t/\pi$, so that

$u = (t - \pi)/\pi$ and $du = dt/\pi$. Equation (5.42) then transforms into

$$P_t = \frac{Z_0 I_0^2}{8\pi} \int_{s=2\pi}^{0} \frac{\cos^2\left(\frac{\pi - s}{2}\right)}{s} d(-s) + \frac{Z_0 I_0^2}{8\pi} \int_{t=0}^{2\pi} \frac{\cos^2\left(\frac{t-\pi}{2}\right)}{t} dt$$

$$= \frac{Z_0 I_0^2}{4\pi} \int_0^{2\pi} \frac{\cos^2\left(\frac{q-\pi}{2}\right)}{q} dq. \tag{5.43}$$

Then, with $\cos^2 A = \frac{1}{2}\cos 2A + \frac{1}{2}$, we get

$$P_t = \frac{Z_0 I_0^2}{8\pi} \int_0^{2\pi} \left[\frac{\cos(q - \pi) + 1}{q} \right] dq, \tag{5.44}$$

and finally, with $\cos(q - \pi) = -\cos q$,

$$P_t = \frac{Z_0 I_0^2}{8\pi} \int_0^{2\pi} \left[\frac{1 - \cos q}{q} \right] dq. \tag{5.45}$$

Now we have the equation we were looking for. In [2] (equation 5.2.2) we find for the *cosine integral* $Ci(z)$

$$Ci(z) = -\int_z^\infty \frac{\cos t}{t} dt = \gamma + \ln z + \int_0^z \frac{\cos t - 1}{t} dt, \tag{5.46}$$

where γ is Euler's constant, $\gamma = 0.5772156649\ldots$ and $Ci(z)$ is tabulated in [2]. Equation (5.45) can now be written as

$$P_t = \frac{Z_0 I_0^2}{8\pi} \left[\gamma + \ln(2\pi) + Ci(2\pi) \right]. \tag{5.47}$$

From the tables in [2] it is found that $Ci(2\pi) \approx -0.02$, so that

$$P_t = \frac{Z_0 I_0^2}{8\pi} \cdot 2.44. \tag{5.48}$$

The directivity function is now found upon substitution of equations (5.37) and (5.48) in equation (5.16):

$$D(\vartheta) = \frac{P(\vartheta)}{P_t/4\pi} = 1.64 \left[\frac{\cos\left(\frac{\pi}{2}\cos\vartheta\right)}{\sin\vartheta} \right]^2, \tag{5.49}$$

and the directivity is found as

$$D = [D(\vartheta)]_{max} = 1.64. \tag{5.50}$$

As we have already concluded from inspection of Figure 5.6, the directivity of a half-wave dipole antenna is slightly larger than that of an elementary dipole antenna ($D = 1.5$).

5.2.1.2 Full-Wavelength Antenna

For $2l = \lambda_0$, equation (5.36) reduces to

$$P(\vartheta) = \frac{Z_0 I_0^2}{8\pi^2} \left[\frac{\cos{(\pi \cos{\vartheta})} + 1}{\sin{\vartheta}} \right]^2. \tag{5.51}$$

The maximum is found for $\vartheta = \frac{\pi}{2}$ and equals $P\left(\frac{\pi}{2}\right) = \frac{Z_0 I_0^2}{2\pi^2}$, so that the normalized radiation pattern is given by

$$F(\vartheta) = \frac{1}{4} \left[\frac{\cos{(\pi \cos{\vartheta})} + 1}{\sin{\vartheta}} \right]^2. \tag{5.52}$$

A polar, linear plot of the radiation pattern in a plane $\varphi = \text{constant}$ is shown in Figure 5.7. In the same figure we also show, for comparison, the radiation pattern cut for an elementary dipole antenna and for a half-wave dipole antenna.

This figure shows that the radiation pattern of a full-wave dipole antenna is similar to those of an elementary and a half-wave dipole antenna, but that the directivity is substantially larger than those of the elementary and half-wave dipole antenna ($D = 1.5$ and $D = 1.64$, respectively).

For calculating the directivity function, we need to know the total radiated power. Therefore we need to integrate equation (5.51) over all angles φ and ϑ and transform the integrals into cosine integrals. For the full-wave dipole antenna this process is, as we will see, even less straight-forward as for the half-wave dipole antenna.

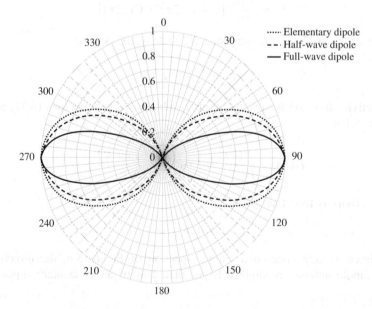

Figure 5.7 Polar radiation pattern cuts in a plane $\varphi = \text{constant}$ on a linear amplitude scale.

The total radiated power is calculated as

$$P_t = \int_{\varphi=0}^{2\pi} \int_{\vartheta=0}^{\pi} P(\vartheta) \sin\vartheta \, d\vartheta \, d\varphi = \frac{Z_0 I_0^2}{4\pi} \int_0^{\pi} \left[\frac{\cos(\pi\cos\vartheta)+1}{\sin\vartheta}\right]^2 \sin\vartheta \, d\vartheta. \quad (5.53)$$

Substitution of $\cos\vartheta = u$ and thus $\sin^2\vartheta = 1 - u^2$ and $\sin\vartheta \, d\vartheta = -du$ in equation (5.53) results in

$$\begin{aligned}
P_t &= \frac{Z_0 I_0^2}{4\pi} \int_{-1}^{1} \frac{[\cos(\pi u) + 1]^2}{1 - u^2} du \\
&= \frac{Z_0 I_0^2}{4\pi} \int_{-1}^{1} \frac{\cos^2(\pi u)}{1 - u^2} du + \frac{Z_0 I_0^2}{4\pi} \int_{-1}^{1} \frac{1}{1 - u^2} du + \frac{Z_0 I_0^2}{4\pi} 2 \int_{-1}^{1} \frac{\cos(\pi u)}{1 - u^2} du \quad (5.54)
\end{aligned}$$

Next, we use $1/(1 - u^2) = 1/(1-u)(1+u) = \frac{1}{2}/(1-u) + \frac{1}{2}/(1+u)$ and obtain

$$\begin{aligned}
P_t &= \frac{Z_0 I_0^2}{8\pi} \int_{-1}^{1} \frac{\cos^2(\pi u)}{1 - u} du + \frac{Z_0 I_0^2}{8\pi} \int_{-1}^{1} \frac{\cos^2(\pi u)}{1 + u} du \\
&\quad + \frac{Z_0 I_0^2}{8\pi} \int_{-1}^{1} \frac{1}{1 - u} du + \frac{Z_0 I_0^2}{8\pi} \int_{-1}^{1} \frac{1}{1 + u} du \\
&\quad + \frac{Z_0 I_0^2}{8\pi} 2 \int_{-1}^{1} \frac{\cos(\pi u)}{1 - u} du + \frac{Z_0 I_0^2}{8\pi} 2 \int_{-1}^{1} \frac{\cos(\pi u)}{1 + u} du. \quad (5.55)
\end{aligned}$$

Then, upon substitution of $1 - u = s/\pi$, so that $u = (\pi - s)/\pi$ and $du = d(-s)/\pi$ and $1 + u = t/\pi$, so that $u = (t - \pi)/\pi$ and $du = dt/\pi$, we get

$$\begin{aligned}
P_t &= \frac{Z_0 I_0^2}{8\pi} \int_0^{2\pi} \frac{\cos^2(s - \pi)}{s} ds + \frac{Z_0 I_0^2}{8\pi} \int_0^{2\pi} \frac{\cos^2(t - \pi)}{t} dt \\
&\quad + \frac{Z_0 I_0^2}{8\pi} \int_0^{2\pi} \frac{1}{s} ds + \frac{Z_0 I_0^2}{8\pi} \int_0^{2\pi} \frac{1}{t} dt \\
&\quad + \frac{Z_0 I_0^2}{8\pi} 2 \int_0^{2\pi} \frac{\cos(s - \pi)}{s} ds + \frac{Z_0 I_0^2}{8\pi} 2 \int_0^{2\pi} \frac{\cos(t - \pi)}{t} dt \\
&= \frac{Z_0 I_0^2}{4\pi} \int_0^{2\pi} \frac{\cos^2(q - \pi)}{q} dq + \frac{Z_0 I_0^2}{4\pi} \int_0^{2\pi} \frac{1}{q} dq + \frac{Z_0 I_0^2}{4\pi} \int_0^{2\pi} \frac{\cos(q - \pi)}{q} dq. \quad (5.56)
\end{aligned}$$

With the substitution of $\cos(q - \pi) = -\cos q$ and $\cos^2 q = \frac{1}{2}\cos 2q + \frac{1}{2}$ in equation (5.56), we get

$$\begin{aligned}
P_t &= \frac{Z_0 I_0^2}{4\pi} \frac{1}{2} \int_0^{2\pi} \frac{1 + \cos 2q}{q} dq + \frac{Z_0 I_0^2}{4\pi} \int_0^{2\pi} \frac{1}{q} dq - \frac{Z_0 I_0^2}{4\pi} 2 \int_0^{2\pi} \frac{\cos q}{q} dq \\
&= \frac{Z_0 I_0^2}{4\pi} \frac{1}{2} \int_0^{2\pi} \frac{1 + \cos 2q}{q} dq + \frac{Z_0 I_0^2}{4\pi} 2 \int_0^{2\pi} \frac{1}{q} dq
\end{aligned}$$

$$- \frac{Z_0 I_0^2}{4\pi} \int_0^{2\pi} \frac{1}{q} dq - \frac{Z_0 I_0^2}{4\pi} 2 \int_0^{2\pi} \frac{\cos q}{q} dq$$

$$= \frac{Z_0 I_0^2}{4\pi} 2 \int_0^{2\pi} \frac{1 - \cos q}{q} dq - \frac{Z_0 I_0^2}{4\pi} \frac{1}{2} \int_0^{2\pi} \frac{1 - \cos 2q}{q} dq, \qquad (5.57)$$

and finally, upon substitution of $2q = v$ in the second integration, we obtain the desired equation:

$$P_t = \frac{Z_0 I_0^2}{4\pi} 2 \int_0^{2\pi} \frac{1 - \cos q}{q} dq - \frac{Z_0 I_0^2}{4\pi} \frac{1}{2} \int_0^{4\pi} \frac{1 - \cos v}{v} dv. \qquad (5.58)$$

With $Cin(z) = \int_0^z \frac{1 + \cos t}{t} dt$ being tabulated in [2], we can write the above equation as

$$P_t = \frac{Z_0 I_0^2}{8\pi^2} [4\pi Cin(2\pi) - \pi Cin(4\pi)]$$

$$= \frac{Z_0 I_0^2}{8\pi^2} [4\pi \cdot 2.4376543 - \pi \cdot 3.1143565] = \frac{Z_0 I_0^2}{8\pi^2} 6.64\pi. \qquad (5.59)$$

Substitution of equations (5.51) and (5.59) in equation (5.16) gives the directivity function

$$D(\vartheta) = \frac{P(\vartheta)}{P - t/4\pi} = \frac{1}{1.66} \left[\frac{\cos(\pi \cos \vartheta) + 1}{\sin \vartheta} \right]^2, \qquad (5.60)$$

and the directivity is found as

$$D = [D(\vartheta)]_{max} = D \left(\vartheta = \frac{\pi}{2} \right) = \frac{4}{1.66} = 2.41. \qquad (5.61)$$

Indeed, the directivity of a full-wave dipole antenna is substantially larger than the directivity of an elementary dipole ($D = 1.5$) and of a half-wave dipole antenna ($D = 1.64$).

5.2.1.3 $\frac{3}{2}$-Wavelength Antenna

For $2l = \frac{3}{2}\lambda_0$, equation (5.36) reduces to

$$P(\vartheta) = \frac{Z_0 I_0^2}{8\pi^2} \left[\frac{\cos \left(\frac{3\pi}{2} \cos \vartheta \right)}{\sin \vartheta} \right]^2. \qquad (5.62)$$

The maximum is found for $\vartheta \approx 43°$. Figure 5.8 shows the normalized radiation pattern in a plane $\varphi = $ constant.

For a dipole antenna having a length of one-and-a-half wavelengths, the radiation is directed mainly in 'conical rings' along the dipole axis. This makes the practical application of such an antenna limited. Therefore, we will not calculate the gain of this antenna.

The reason we have shown the radiation pattern cut is to demonstrate the effect of having a dipole length exceeding one wavelength, that is the appearance of 'elevational

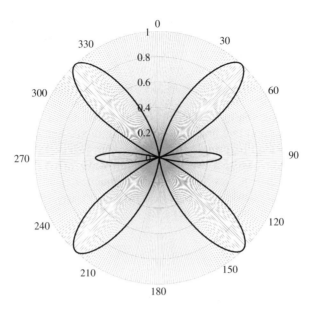

Figure 5.8 Polar radiation pattern cut in a plane $\varphi =$ constant on a linear amplitude scale for a $\frac{3}{2}$-wavelength dipole antenna.

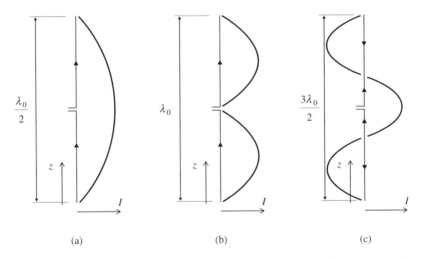

Figure 5.9 Current densities over wire antennas of different lengths. (a) Half-wave dipole antenna. (b) Full-wave dipole antenna. (c) $\frac{3}{2}$-wave dipole antenna.

lobes' in the radiation pattern. To explain this phenomenon, we have to take a closer look at the current density along the wire antenna. This is done in Figure 5.9 for a number of antenna lengths.

We know that the source of electromagnetic radiation is acceleration or deceleration of charge. In Figure 5.9(a) and (b), sources of radiation are the dipole excitation point and the two end-points. The figure shows that the current densities at these source points are all in phase and thus add up to a contribution (radiation) perpendicular to the wire axis.

In Figure 5.9(c) the antenna length has exceeded a wavelength and additional sources (areas of acceleration or deceleration) are created. Now, all the sources are *not* in phase and a more complicated pattern of directions of constructive and destructive interference is created, resulting in the creation of elevational lobes, see Figure 5.8. We will learn more about this phenomenon in Chapter 8 where we will discuss array antennas.

We have to bear in mind that antennas – including dipole antennas – are always used for frequency bands rather than single frequencies. This means that we have to be careful in choosing the dipole length. If we design a full-wave dipole, we must prevent the length getting larger than a wavelength in the frequency band being used. If we fix the length for the longest wavelength (smallest frequency), we will see the appearance of elevational lobes at the highest frequency.

5.2.2   Input Impedance

We have seen – in discussing the elementary dipole – that, based on the total transmitted power, we can calculate the radiation resistance. In doing so we neglect the antenna impedance having a reactive part. For a more accurate calculation of the antenna input impedance, the current distribution needs to be known more accurately than the, for now, assumed sinusoidal distribution. The calculation of an accurate, complex dipole antenna input impedance is, however, beyond the scope of this book (and most antenna textbooks). For antenna lengths not exceeding a wavelength, the real part of the input impedance will be close to the radiation resistance as calculated from the total radiated power. For antenna lengths in the range of half a wavelength, excellent approximate, closed-form equations do exist [3].

We will derive the radiation resistance for a half-wave dipole antenna and for a full-wave dipole antenna. Due to the limited practical use of a $\frac{3}{2}$-wave dipole antenna we will not calculate its radiation resistance. The radiation resistance will be (almost) equal to the real part of the antenna input impedance.

5.2.2.1   Radiation Resistance Half-Wavelength Antenna

We find the radiation resistance by substituting the total radiated power in equation (5.21). So for the half-wave dipole antenna, the radiation resistance is found to be

$$R_A = \frac{P_t}{\frac{1}{2}I_0^2} = \frac{\frac{Z_0 I_0^2}{8\pi} \cdot 2.44}{\frac{1}{2}I_0^2} = \frac{2.44 \cdot 120\pi}{4\pi} = 73.2\,\Omega, \qquad (5.63)$$

since $Z_0 = 120\pi\,\Omega$.

This antenna impedance is close to the 75 Ω of the standard RG-6 coaxial cable. The reflection coefficient upon connecting a half-wave dipole antenna to a RG-6 coaxial cable would be 0.012, which is negligible. So with the half-wave dipole antenna we have maintained the favorable near-omnidirectional radiation characteristics of an elementary dipole antenna but we have also created an input impedance that allows an easy application of the antenna in practice. As said before, the half-wavelength dipole antenna is a very much used antenna. The reasons therefore have just been given.

5.2.2.2 Input Impedance Half-Wavelength Antenna

The input impedance of a wire antenna may be calculated using the induced EMF method. A fitting method applied to a first fitting derived by C.T. Tai gives, for dipole lengths around half a wavelength [3],

$$Z_{in} = \left[122.65 - 204.1k_0l + 110\,(k_0l)^2\right]$$

$$- j\left[120\left(\ln\frac{2l}{a} - 1\right)\cot k_0l - 162.5 + 140k_0l - 40\,(k_0l)^2\right]$$

$$\text{for } (1.3 \le k_0l \le 1.7) \text{ and } (0.001588 \le a/\lambda_0 \le 0.009525)\,, \tag{5.64}$$

where a is the radius of the wire.

For $2l = \lambda_0/2$ and $a = 0.005\lambda_0$, equation (5.64) gives $Z_{in} = (73.46 - j15735)\,\Omega$. We see that the real part of the input impedance is indeed well approximated by our earlier radiation resistance calculation.

5.2.2.3 Radiation Resistance Full-Wavelength Antenna

The real part of the input impedance for a center-fed, full-wave dipole antenna is infinite if the wire is perfectly electrically conducting or very large for a non-perfectly-conducting wire, since the current density is (approaching) zero at the center, see Figure 5.9(b). By feeding at a current density maximum, see Figure 5.10, we can calculate the radiation resistance by substituting equation (5.59) in equation (5.21)

$$R_A = \frac{P_t}{\frac{1}{2}I_0^2} = \frac{\frac{Z_0 I_0^2}{8\pi} \cdot 6.64}{\frac{1}{2}I_0^2} = 199.2\,\Omega, \tag{5.65}$$

5.3 Printed Monopole and Inverted-F Antennas

In this section we will be using the commercially available software package Microwave Studio® from Computer Simulation Technology [4], CST-MWS. CST-MWS is a specialist tool for the three-dimensional electromagnetic simulation of high frequency (HF) components. It enables the fast and accurate analysis of high frequency devices such as antennas, filters, couplers, planar and multi-layer structures, and system integrity (SI) and electromagnetic compatibility (EMC) effects. The software allows us to switch between different important methods for antenna simulation, among others the finite integration technique (FIT), including time domain, finite element method (FEM) and method of moments (MoM). We will be using the FIT transient and frequency domain solvers for designing printed monopole antennas and for designing an inverted-F antenna (IFA).

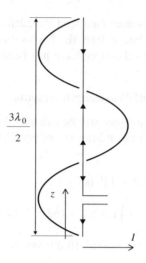

Figure 5.10   Current densities over full-wave, wire antenna, excited at a current density maximum.

5.3.1   Application of Theory

We will now apply what we have learnt about dipole antennas to design practical dipole-like antennas. First we will design a printed monopole antenna followed by the design of a so-called inverted-F antenna (IFA). Then we will design an ultra wideband (UWB) monopole antenna and a miniature monopole antenna with cable current suppression means. All antennas will be made in microstrip or coplanar waveguide (CPW) technology and may be integrated on a printed circuit board (PCB) of a mobile communications device.

We will not perform the designs by going into the mathematical details. Instead, we will use the found relationships between dipole antenna dimensions and characteristics to input an initial design into CST-MWS. Through analyzing this design and – based on the analysis results and the mentioned relationships – fine tuning some dimensions, we will obtain our final design in a limited number of iterations.

The basis of the printed monopole antennas and the inverted-F antenna is the monopole antenna. A monopole antenna is one half of a dipole antenna, placed above a perfect electric conducting (PEC) ground plane, see Figure 5.11(a).

The fields above the PEC ground may be found from the equivalent dipole antenna in free space shown in Figure 5.11(b). This equivalent dipole antenna consists of the original monopole antenna and its image in the PEC ground plane. For a quarter-wave ($l = \frac{\lambda}{4}$) monopole antenna, the fields above the ground plane are identical to those of the free space half-wavelength ($2l = \frac{\lambda}{2}$) dipole antenna. The input impedance of the quarter-wave monopole antenna is half that of the half-wave dipole antenna. The directivity of the monopole antenna is twice that of the dipole antenna.

Quarter-wave monopole antennas used to be popular in previous generations of mobile phones; nowadays IFA and planar IFA (PIFA) structures are being employed.

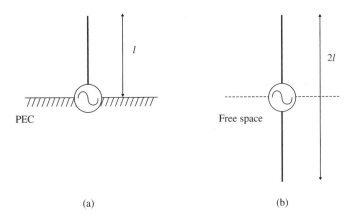

(a)   (b)

Figure 5.11   Monopole antenna. (a) Monopole over PEC. (b) Equivalent dipole in free space.

5.3.2   *Planar Monopole Antenna Design*

For a practical implementation of the monopole antenna, we first need to replace the infinite ground plane by a finite one. Then, for integration in a PCB in microstrip technology, we need to transfer the wire monopole that is perpendicular to the (now) finite ground plane into a planar structure as shown in Figure 1.9(a) and Figure 5.12(a).

The strip monopole antenna is an extension of a $50\,\Omega$ microstrip transmission line beyond the rim of the ground plane. The excitation of the monopole antenna is at the rim, see Figure 5.12(b).

Another way of viewing this printed monopole antenna is regarding the structure as an asymmetrically driven strip dipole antenna. The strip monopole antenna is one of the arms

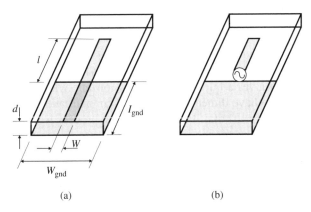

(a)   (b)

Figure 5.12   Printed monopole antenna. (a) Microstrip excited printed monopole antenna. (b) Discrete port excited printed monopole antenna.

of the asymmetrically driven strip dipole antenna, the ground plane forming the other arm. Therefore, the ground plane length should not exceed a quarter of a wavelength at the desired resonance frequency. The length of the strip monopole should be a quarter of a wavelength at the desired resonance frequency. Although the dielectric of the PCB does influence the wavelength, the free space quarter wavelength will be a good starting point in designing the printed monopole antenna.

The width of the strip will be dictated by the microstrip transmission line. We want this transmission line to have a characteristic impedance of 50 Ω. If we fine tune the monopole antenna to having an input impedance of 50 Ω at the desired frequency, the length of the microstrip transmission line will have no influence on the antenna characteristics and will not form part of the antenna.

To design a microstrip transmission line of a desired characteristic impedance, we choose the width of the strip W as [1]

$$
\frac{W}{d} = \begin{cases} \frac{8e^A}{e^{2A}-2} & \text{if } \frac{W}{d} < 2, \\[2ex] \frac{2}{\pi}\left[B-1-\ln(2B-1)+\frac{\varepsilon_r-1}{2\varepsilon_r}\left\{\ln(B-1)+0.39-\frac{0.61}{\varepsilon_r}\right\}\right] & \text{if } \frac{W}{d} > 2, \end{cases}
$$

$$(5.66)$$

where d is the thickness of the dielectric, ε_r is the relative permittivity of the dielectric and [1]

$$
A = \frac{Z_0}{60}\sqrt{\frac{\varepsilon_r+1}{2}} + \frac{\varepsilon_r-1}{\varepsilon_r+1}\left(0.23+\frac{0.11}{\varepsilon_r}\right), \qquad (5.67)
$$

$$
B = \frac{377\pi}{2Z_0\sqrt{\varepsilon_r}}. \qquad (5.68)
$$

Z_0 in equations (5.67) and (5.68) is the characteristic impedance of the microstrip transmission line. Transmission line theory is explained in Appendix F.

We now start with the design of a printed monopole antenna, resonant at a frequency of 2.45 GHz.

The PCB material will be 1.6 mm thick FR4 having a relative permittivity $\varepsilon_r = 4.28$ and a loss tangent $\tan\delta = 0.016$. Substituting $Z_0 = 50\,\Omega$, d = 1.6 mm and $\varepsilon_r = 4.28$ in equations (5.66)–(5.68) gives a microstrip transmission line width of W = 3.1 mm.

We choose the PCB to have dimensions 8 cm × 4 cm and take for the initial length of the strip monopole $l = \frac{\lambda_0}{4} = 31$ mm. In the fine tuning process we will probably need to shorten this length to account for the wavelength decrease due to the dielectric. For the reasons mentioned before we choose the length of the ground plane l_{gnd} to be equal to 30 mm. The metal layers will be copper with a thickness of 70 μm.

After launching CST-MWS, see Figure 5.13, and choosing an appropriate template, that, among others, sets the correct boundary conditions (for this example, we opt for *antenna, mobile phone*, see Figure 5.14), we input the antenna structure, see Figure 5.15. We select the frequency range and set field monitors, see Figure 5.16 and perform an adaptive (frequency domain) analysis.

The reflection coefficient as a function of frequency is shown in Figure 5.17 and the radiation pattern at 2.45 GHz is shown in Figure 5.18.

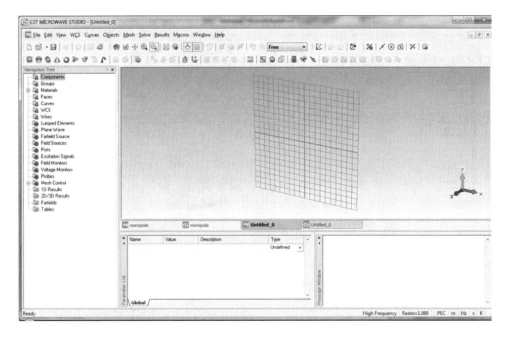

Figure 5.13   Drawing a printed monopole antenna in CST-MWS. Launching CST-MWS.

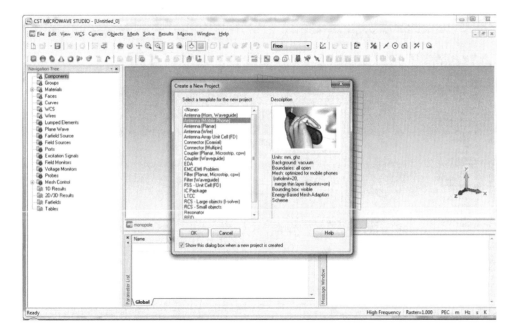

Figure 5.14   Drawing a printed monopole antenna in CST-MWS. Choosing a template.

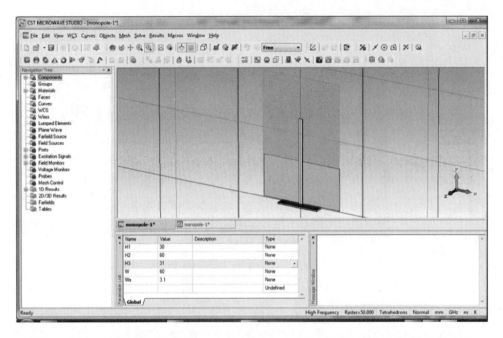

Figure 5.15 Drawing a printed monopole antenna in CST-MWS. Drawing the structure.

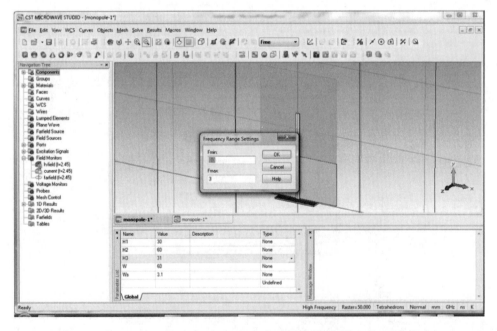

Figure 5.16 Drawing a printed monopole antenna in CST-MWS. Choosing the frequency range and setting the field monitors.

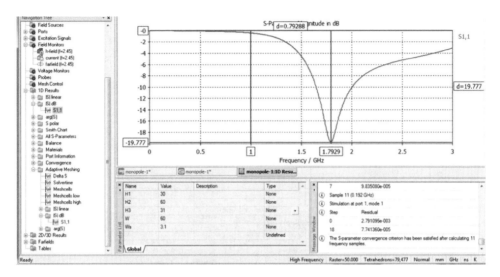

Figure 5.17   Analysis results of initial printed monopole antenna design. Reflection amplitude as a function of frequency.

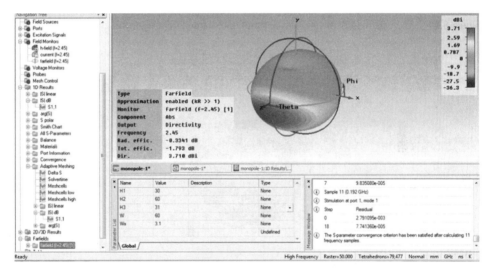

Figure 5.18   Analysis results of initial printed monopole antenna design. Radiation pattern at 2.45 GHz.

Figures 5.17 and 5.18 show that it is behaving like a dipole antenna and that the antenna is resonant at a frequency $f = 1.79$ GHz. The resonance frequency is too low. Therefore we need to decrease the monopole length and possibly the ground plane length also.

After a few iterations we find that for a dipole length of $l = 21$ mm and a ground plane length of $l_{gnd} = 20$ mm, we get a resonance and dipole antenna behavior at 2.45 GHz as desired, see Figures 5.19 and 5.20.

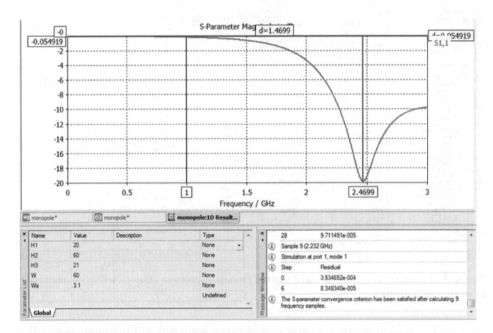

Figure 5.19 Analysis results of final printed monopole antenna design. Reflection amplitude as a function of frequency.

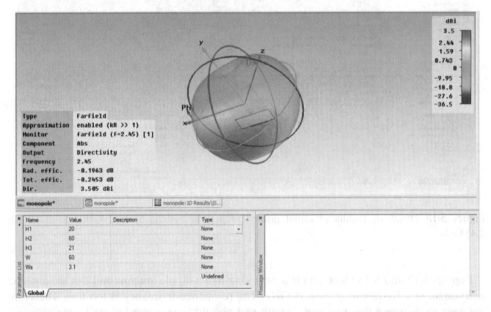

Figure 5.20 Analysis results of final printed monopole antenna design. Radiation pattern at 2.45 GHz.

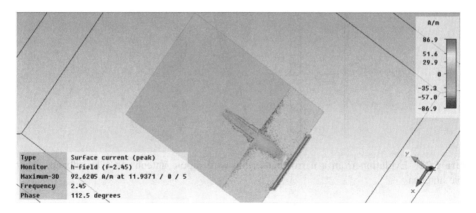

Figure 5.21   Current density analysis results of the final printed monopole antenna.

Finally, Figure 5.21 shows the surface current density on the antenna structure. The Figure shows that the ground plane is wide enough to avoid the creation of additional radiation from the corners.

5.3.3   Printed UWB Antenna Design

High data rate wireless communications need wide frequency bandwidths. In the ultra wideband (UWB) frequency band from 3.1 GHz to 10.6 GHz, information may be spread over a large bandwidth at low power levels, thus creating the possibility of sharing the spectrum with other users. The dipole antenna offers a good starting point for developing a compact antenna having an impedance bandwidth, wide enough to cover the whole frequency band from 3.1 GHz to 10.6 GHz.

By thickening the arms of a dipole antenna, current may travel over different paths of different lengths, thus increasing the bandwidth. The current distribution is then – unlike for a thin dipole antenna – no longer sinusoidal. While this hardly affects the radiation pattern of the antenna, it severely influences the input impedance [5]. The band-widening effect may be further exploited by shaping the thick dipole arms into cones or spheres [6]. Figure 5.22 shows the evolution from a thin-wire dipole antenna to a spherical-arm dipole antenna [7].

For practical applications we will transform the dipole with two spherical arms into a planar version that may be integrated into or onto a printed circuit board (PCB). Based on a planar stripline excited UWB antenna [8], a microstrip excited pseudo-monopole as shown in Figure 5.23 has been devised [7].

The antenna is basically a dipole antenna consisting of two planar, circular arms.

The upper arm, on the top side of the PCB, is connected to the microstrip transmission line trace. The lower arm, on the underside of the PCB is integrated into a small ground area, together providing the ground plane for the microstrip trace.

The functioning of the antenna may be explained as being a dipole antenna for the frequency for which the antenna is half a wavelength long and as being a dual tapered slot antenna for higher frequencies. The latter is obvious from the current densities as

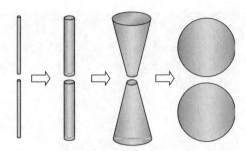

Figure 5.22 Evolution from a narrowband, thin-wire dipole antenna to a broadband, spherical dipole antenna.

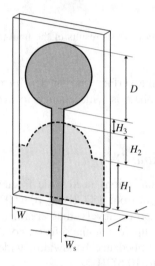

Figure 5.23 Microstrip excited UWB antenna.

shown in Figures 5.24 and 5.25. Figure 5.24 shows the surface current density at 3 GHz for a planar circular dipole antenna resonant around that frequency.

We see that current is flowing along the whole metallic perimeter of the antenna, having a maximum in between the discs and a minimum on top of the upper disc and at the bottom of the ground plane. The antenna thus acts as a half-wave dipole antenna.

The surface current density for the same antenna structure, analyzed at 7 GHz is shown in Figure 5.25. Again we see the current density concentrated at the rims of the discs, but now the current attenuates when traveling over the rim. Now the antenna structure may be seen as consisting of two flared notch antennas, see Figure 5.26.

The equivalence is best for the current density disappearing at the indicated positions. The radiation of the two equivalent flared notch antennas (of which one is shown in Figure 5.26) is in the same direction as that of the equivalent half-wave dipole antenna at lower frequencies. For increasing frequencies, the flared notch antenna equivalence still holds, but the current density attenuates 'faster', going over the rims of the discs.

So, the UWB antenna starts at the lower frequency bound as a half-wave dipole antenna and transforms into a dual flared notch antenna going to higher frequencies.

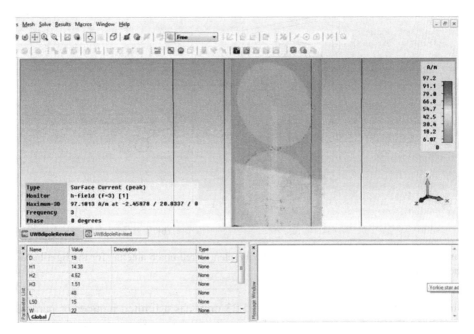

Figure 5.24   Surface current density at 3 GHz for a 19 mm diameter printed circular dipole antenna.

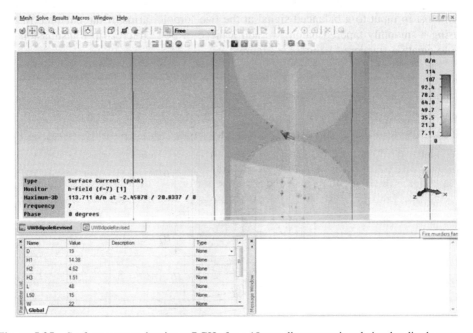

Figure 5.25   Surface current density at 7 GHz for a 19 mm diameter printed circular dipole antenna.

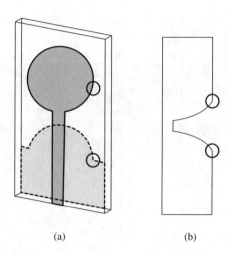

(a)   (b)

Figure 5.26   Printed UWB antenna (a) and equivalent flared notch antenna for the right-hand side of the structure (b).

Since the structure as shown in Figure 5.23 only has a limited number of parameters to tune once the starting frequency is known, it will be necessary to use the microstrip transmission line width in the structure to tune the antenna input impedance. As a consequence, this section of microstrip transmission line has become part of the antenna. A better solution is provided by using a microstrip balun to transfer the unbalanced signal at the microstrip input to a balanced signal at the two 'dipole' arms of the antenna. Instead of using a smoothly tapered balun, we add a piece of 50 Ω microstrip transmission line (H_4) to create a staggered balun, see Figure 5.27.

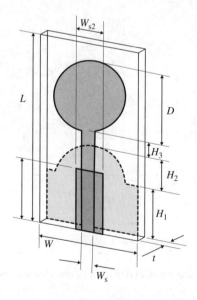

Figure 5.27   Printed UWB antenna with staggered balun.

Through varying the different dimensions, we obtain satisfactory results for the dimensions and parameters stated in Table 5.1. The thickness of the copper is 70 μm.

The structure is shown in Figure 5.28. The reflection as a function of frequency is shown in Figure 5.29.

Figures 5.30–5.33 show the radiation patterns at 3 GHz, 6 GHz, 7 GHz and 10 GHz, respectively.

Table 5.1 Dimensions and parameters of printed UWB antenna

Dimension/Parameter	Value
D	19 mm
H_1	14.38 mm
H_2	4.62 mm
H_3	1.51 mm
H_4	15.0 mm
L	48.0 mm
W	22.0 mm
W_s	1.44 mm
W_{s2}	3.1 mm
t	1.6 mm
ε_r	4.28
$\tan \delta$	0.016

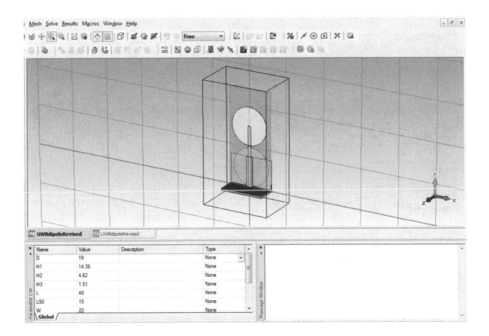

Figure 5.28 Printed UWB antenna.

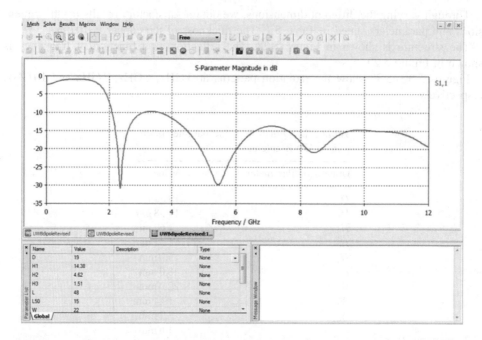

Figure 5.29   Reflection as a function of frequency for the antenna shown in Figure 5.28.

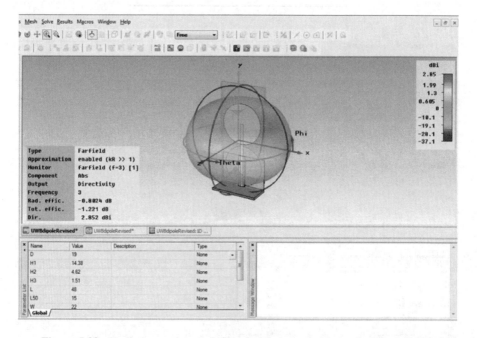

Figure 5.30   Radiation pattern at 3 GHz for the antenna shown in Figure 5.28.

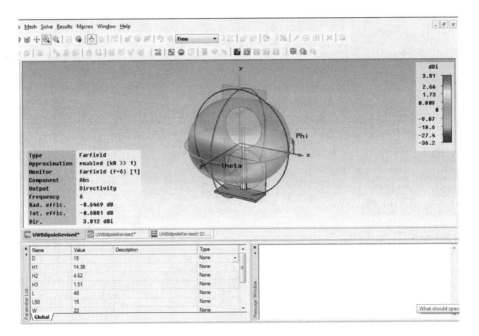

Figure 5.31 Radiation pattern at 6 GHz for the antenna shown in Figure 5.28.

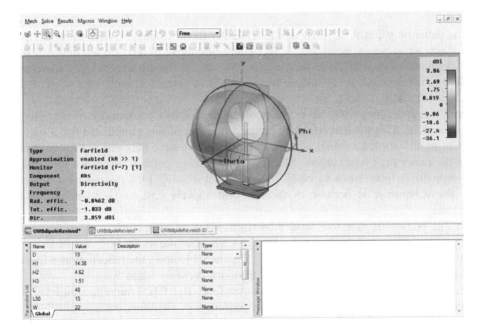

Figure 5.32 Radiation pattern at 7 GHz for the antenna shown in Figure 5.28.

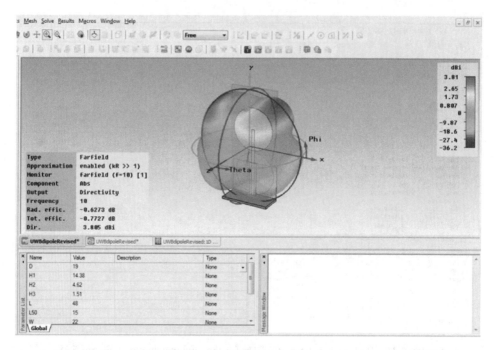

Figure 5.33   Radiation pattern at 10 GHz for the antenna shown in Figure 5.28.

The radiation patterns at 3 GHz and to a slightly lesser extent at 6 GHz show a 'dipole-like' behavior as was to be expected. At frequencies above 6 GHz, we see the occurrence of elevational lobes. For those frequencies, the 'dipole' length exceeds one wavelength leading to the occurrence of multiple lobes. This is in agreement with what we have seen in Section 5.2.1.3.

In the analysis shown we have assumed that the relative permittivity of the substrate was $\varepsilon_r = 4.28$. This value is based on the measurement of one sample of FR4, a standard PCB material.[5] FR4, however, is not a dedicated microwave laminate, which is obvious from the rather high loss tangent. Moreover, the relative permittivity is not well defined and may differ from batch to batch. The reason for using FR4 instead of a dedicated microwave laminate is because of the low cost and the ease of photo-etching copper structures.

At this point it is therefore wise to perform a tolerance analysis with respect to the relative permittivity of the substrate. Within CST Microwave Studio® we make use of the property to perform a parameter sweep. Figure 5.34 shows the reflection as a function of frequency for values of the relative permittivity between 3 and 5.

We see that in the frequency band of interest (3.1 – 10.6 GHz) the reflection level stays under −10 dB for all values of the relative permittivity. Therefore we may conclude that it is permitted to make this antenna on FR4 regardless of its not-well-specified relative permittivity.

---

[5] A list of relative permittivities for different materials is given in Appendix D.

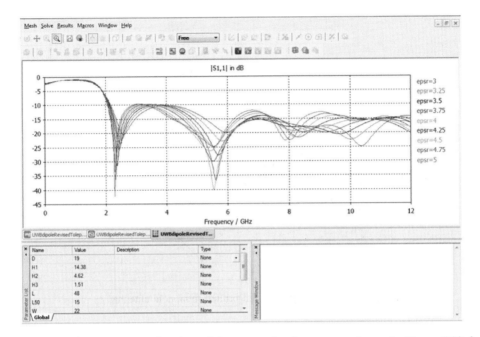

Figure 5.34 Reflection as a function of frequency for the antenna shown in Figure 5.28 for different values of the substrate's relative permittivity.

5.3.4 Miniature Monopole with Cable Current Suppression

In the next design case we will look at the influence of a coaxial cable connected to a small monopole antenna and means to suppress this influence.

We start with a coplanar waveguide (CPW) excited monopole antenna on FR4, see Figure 5.35.

A coplanar waveguide consists of a metal strip, having width W, in between two insulated ground planes, separated a distance S, on top of a dielectric sheet, see Figure 5.35 and Figure 5.36 for a cross-sectional view.

The advantages of CPW are that (active) devices can be mounted on top of the circuit (as in microstrip) and that only one side of the dielectric sheet needs to be patterned. An extensive treatment of CPW transmission lines is beyond the scope of this book. Analytical equations for the characteristic impedance can be found in Appendix G.

The dimensions and parameters of a CPW excited printed monopole antenna, resonant at 2.45 GHz are stated in Table 5.2, with reference to Figure 5.35. The thickness of the copper layer is 70 µm.

Next, a coaxial cable is attached to the antenna, as shown in Figure 5.37. A copper block is attached to the rim of the PCB, mimicking a sub miniature A (SMA) connector. The relevant dimensions and parameters are shown in Figure 5.38.

The radii of the coaxial cable are, see Figure 5.38(a), $a = 0.375$ mm, $b_1 = 1.5$ mm and $b_2 = 1.75$ mm. The 'SMA mimicking' block dimensions are, see Figure 5.38(b), $W_S = 6.0$ mm and $d_S = 2.0$ mm. In Figure 5.39 we show the reflection coefficient as a function of frequency for different lengths of the connected coaxial cable.

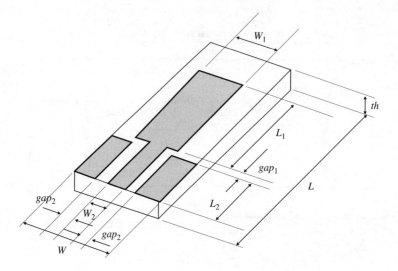

Figure 5.35 CPW excited printed monopole antenna.

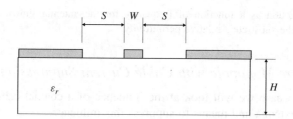

Figure 5.36 Coplanar waveguide, cross-sectional view.

Table 5.2 Dimensions and parameters of CPW
excited printed monopole antenna

Dimension/Parameter	Value
L	48.0 mm
L_1	27.0 mm
L_2	15.0 mm
W	10.0 mm
W_1	6.0 mm
W_2	2.0 mm
gap_1	1.0 mm
gap_2	1.0 mm
th	1.6 mm
ε_r	4.28
$\tan \delta$	0.016

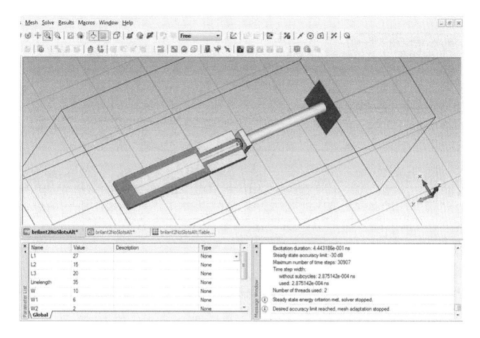

Figure 5.37   CPW excited printed monopole antenna with coaxial cable attached.

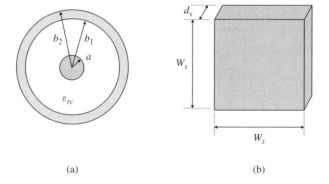

(a)   (b)

Figure 5.38   Cross section of coaxial waveguide (a) and SMA-mimicking copper block (b).

The figure shows that the resonance frequency of the structure decreases with increasing coaxial cable length. This phenomenon can be explained by taking a closer look at the current densities over the metal surfaces of the structure. As shown in Figure 5.40, currents flowing on the CPW ground planes of the printed monopole antenna will continue at the outer surface of the coaxial cable's outer conductor.

Next to these currents, a part of the return current flowing on the inside of the coaxial cable's outer conductor passes through the CPW ground planes and will also flow on the outside surface of the coaxial cable's outer conductor. These two current contributions

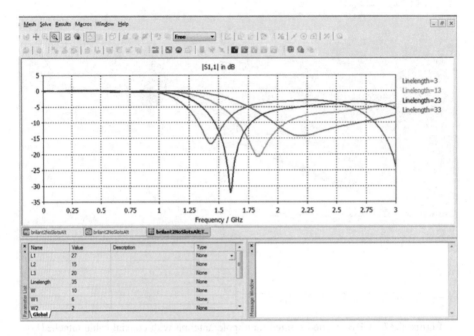

Figure 5.39   Reflection as a function of frequency for different coaxial cable lengths.

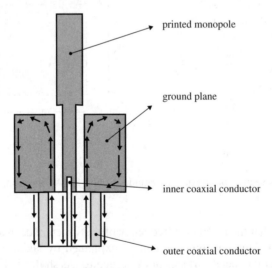

Figure 5.40   Current flowing on the metal surfaces of a printed monopole antenna and a connected coaxial cable.

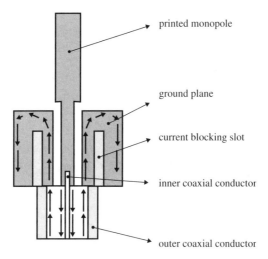

printed monopole

ground plane

current blocking slot

inner coaxial conductor

outer coaxial conductor

Figure 5.41 Current flowing on the metal surfaces of a printed monopole antenna with current blocking slots and a connected coaxial cable.

will add to the radiated field and will influence both the radiation pattern and the input impedance if they are strong enough. So, the well-known shielding effect of the outer conductor of a coaxial cable is made less effective by connecting the cable to the printed monopole antenna.

A means of suppressing these coaxial cable current effects can consist of positioning the connecting coaxial cable perpendicular to the printed monopole antenna, so that the radiation by the cable currents will be in the orthogonal (cross) polarization. Another means is to embed (part) of the coaxial cable in a field-absorbing material to suppress the cable current radiation contribution. Both of these means, however, suppress the effects of cable current instead of suppressing the source itself. A means of suppressing the current flow on the coaxial cable's outer surface is to introduce 'current blocking slots' in the CPW ground planes as shown in Figure 5.41.

The current blocking slots prevent the current passing from the ground planes to the outer surface of the coaxial cable's outer conductor.

The structure as shown in Figure 5.35 now changes into the one shown in Figure 5.42.

The current blocking slot positions and widths must be such that a direct galvanic contact between the CPW ground planes and the outer surface of the coaxial cable's outer conductor is avoided. Within these constraints the structure is optimized for resonance at 2.45 GHz by iteratively changing the various antenna dimensions using CST Microwave Studio®, starting with a quarter free-space wavelength monopole length. The final results are given in Table 5.3.

The structure as analyzed in CST Microwave Studio® is shown in Figure 5.43. In Figure 5.44 the simulated reflection as a function of frequency is shown for several coaxial cable lengths.

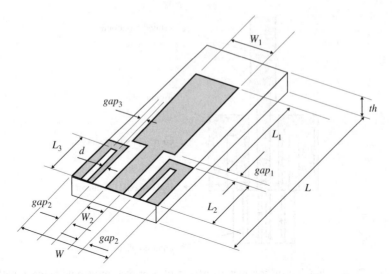

Figure 5.42 CPW excited printed monopole antenna with current blocking slots in the ground planes.

Table 5.3 Dimensions and parameters of CPW excited printed monopole antenna with current blocking slots

Dimension/Parameter	Value
L	66.0 mm
L_1	35.0 mm
L_2	25.0 mm
L_3	20.0 mm
W	10.0 mm
W_1	6.0 mm
W_2	2.0 mm
gap_1	1.0 mm
gap_2	1.0 mm
gap_3	1.0 mm
d	1.0 mm
th	1.6 mm
ε_r	4.28
$\tan \delta$	0.016

The figure shows how the current blocking slots prevent current passing from the CPW ground planes to the outer surface of the coaxial cable's outer conductor. The length of the coaxial cable no longer influences the input impedance and thus the reflection as a function of frequency remains more or less the same for different lengths of coaxial cable.

For the structure without the blocking slots, the coaxial cable becomes part of the antenna. In fact, in that situation we are dealing with an asymmetrically excited dipole antenna. The printed monopole is one of the asymmetric dipole arms, the ground with the attached coaxial cable forms the other arm. In agreement with this reasoning we see

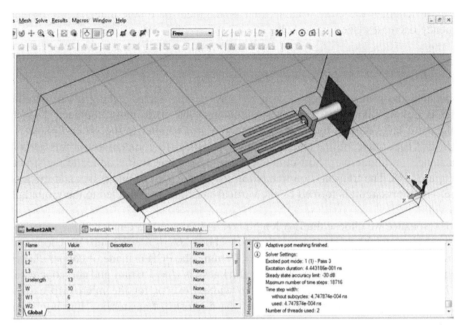

Figure 5.43 CPW excited printed monopole antenna with current blocking slots with attached coaxial cable.

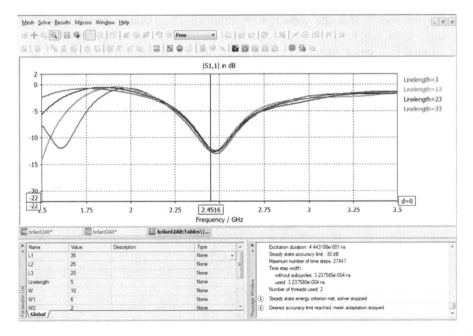

Figure 5.44 Reflection as a function of frequency for the structure shown in Figure 5.43 for several lengths of coaxial cable.

that if we lengthen the coaxial cable, that is lengthen the dipole antenna, the resonance frequency decreases, see Figure 5.39.

5.3.5 Inverted-F Antenna Design

By bending the printed monopole arm by a 90 degree angle, making it run parallel to the PCB ground plane, the antenna structure may be made more compact. The effect of placing a large part of the printed monopole arm parallel to the ground plane is the addition of capacitance to the antenna input impedance. To compensate for this additional capacitance, an (inductive) short-circuited transmission line stub is added to the structure, see Figure 5.45. The transmission line is formed by the strip of width W_{s2} and the ground plane. The short circuit is formed by the vertical strip that is connected to the ground plane by a via.

For resonance we choose the IFA length ($H_3 + L$) equal to a quarter of the wavelength. For the design of a 900 MHz IFA, we therefore choose $H_3 + L = 83$ mm. We choose, more or less arbitrarily, $H_3 = 8$ mm, so that $L = 75$ mm. The PCB is made of copper clad FR4, having $\varepsilon_r = 4.28$ and $\tan \delta = 0.016$. The thickness is $d = 1.6$ mm and the copper traces are 70 μm thick. For convenience all strip widths, except for the microstrip excitation line, are chosen to be equal. The width of the microstrip transmission line is chosen as $W_s = 3.1$ mm, the width for creating a 50 Ω characteristic impedance. These and the other dimensions and parameters of the initial IFA design are shown in Table 5.4.

The structure is shown in Figure 5.46.

The reflection as a function of frequency is shown in Figure 5.47.

The figure shows that the reflection is minimal at around 900 MHz, but that the impedance level can be improved. A perfect, real, near 50 Ω input impedance would have resulted in a much lower reflection at the desired resonance frequency. Therefore

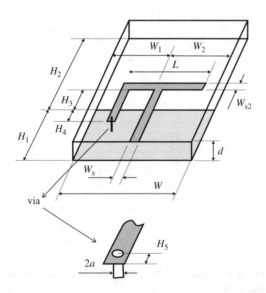

Figure 5.45 Layout of an inverted-F antenna (IFA) in microstrip technology.

Table 5.4 Dimensions and parameters of initial microstrip excited IFA design

Dimension/Parameter	Value
L	75.0 mm
H_1	40.0 mm
H_2	21.0 mm
H_3	8.0 mm
H_4	6.0 mm
H_5	2.0 mm
W_1	40.0 mm
W_2	73.0 mm
W_s	3.1 mm
W_{s2}	3.0 mm
a	1.0 mm
d	1.6 mm
ε_r	4.28
$\tan \delta$	0.016

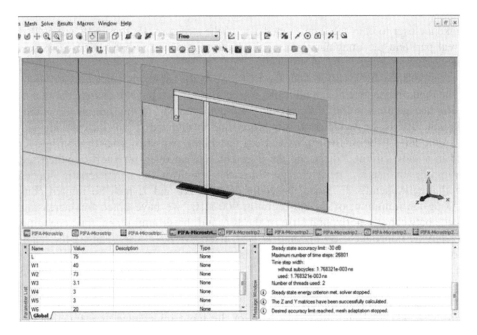

Figure 5.46 Microstrip excited inverted F antenna (IFA). Initial design.

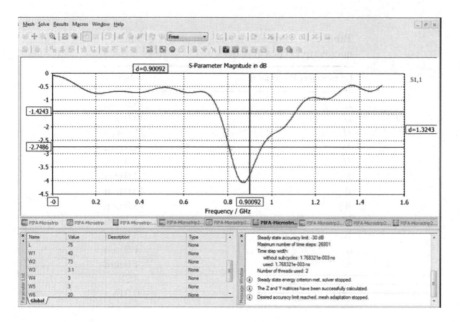

Figure 5.47   Reflection as a function of frequency for the antenna shown in Figure 5.46.

we take a closer look at the input impedance of the structure. Figures 5.48 and 5.49 show the real part and the imaginary part of the input impedance as a function of frequency, respectively.[6]

The impedance curves show that the real part of the input impedance is too high and that the input impedance is inductive. To correct for this, we have to change the two transmission line sections in the IFA structure. These are the part to the right of the microstrip excitation, that is the two-strip transmission line terminated in an open circuit and the part to the left of the microstrip excitation, that is the two-strip transmission line terminated in a short circuit. To change the impedance of an open- or short-circuited section of transmission line we must change either the length or the characteristic impedance or both, see Appendix F. Since the added lengths H_3 and L have to remain constant for fixing the minimum of the reflection at 900 MHz, we have opted for increasing the distance between the horizontal strip of the IFA and the ground plane, thus changing the characteristic impedance. By keeping $H_3 + L$ fixed, we have also changed the length of the open-circuited two-strip transmission line. After a few iterations, we have come up with the structure shown in Figure 5.50. The dimensions, with reference to Figure 5.45, are given in Table 5.5.

The simulated reflection as a function of frequency for this structure is shown in Figure 5.51. The real and imaginary parts of the input impedance as a function of frequency are shown in Figures 5.52 and 5.53, respectively.

---

[6] An experienced, or what we might now call an 'old school', microwave engineer would have used a Smith chart here. The discussion of the Smith chart for this one example only would be too cumbersome. Therefore we look at the real and imaginary parts of the input impedance as a function of frequency separately here. For a discussion of the Smith chart, the reader is referred to [1].

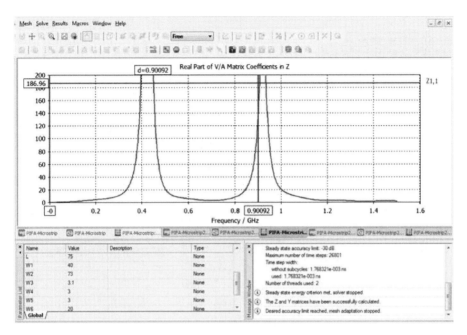

Figure 5.48 Real part of the input impedance as a function of frequency for the antenna shown in Figure 5.46.

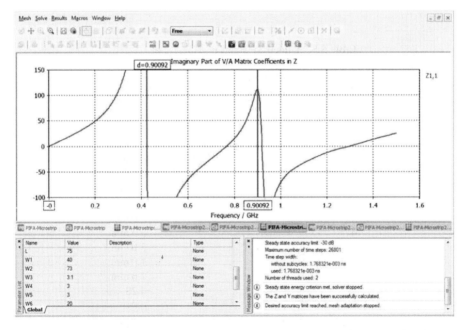

Figure 5.49 Imaginary part of the input impedance as a function of frequency for the antenna shown in Figure 5.46.

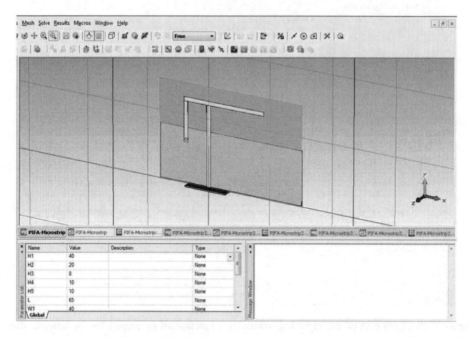

Figure 5.50   Microstrip excited IFA. Redesign.

Table 5.5   Dimensions and parameters of
microstrip excited IFA redesign

Dimension/Parameter	Value
L	65.0 mm
H_1	40.0 mm
H_2	33.0 mm
H_3	20.0 mm
H_4	10.0 mm
H_5	2.0 mm
W_1	40.0 mm
W_2	75.0 mm
W_s	3.3 mm
W_{s2}	3.0 mm
a	1.0 mm
d	1.6 mm
ε_r	4.28
$\tan \delta$	0.016

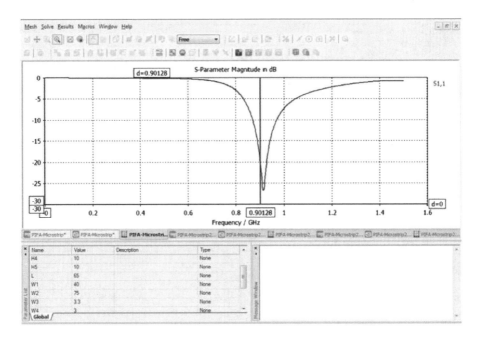

Figure 5.51 Reflection as a function of frequency for the IFA shown in Figure 5.50.

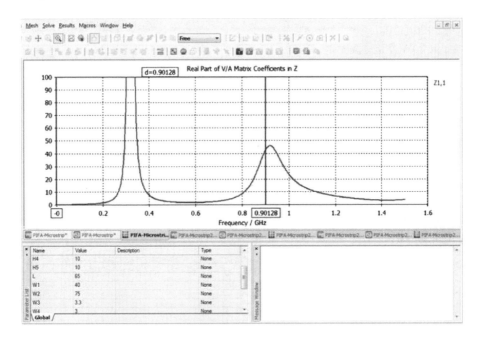

Figure 5.52 Real part of the input impedance as a function of frequency for the IFA shown in Figure 5.50.

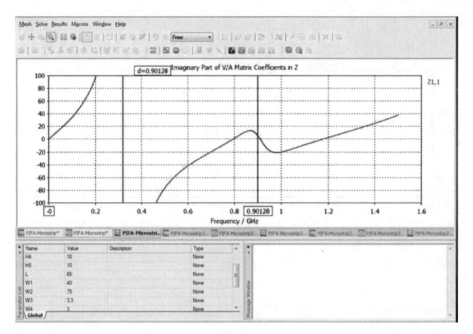

Figure 5.53 Imaginary part of the input impedance as a function of frequency for the IFA shown in Figure 5.50.

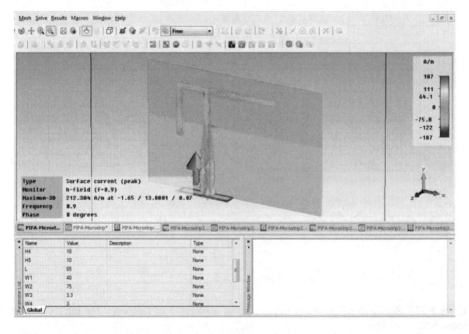

Figure 5.54 Surface current density at 900 MHz for the IFA shown in Figure 5.50.

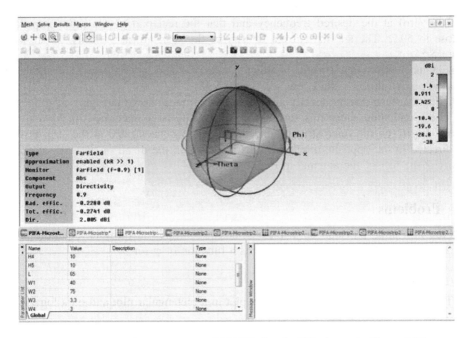

Figure 5.55 Radiation pattern at 900 MHz for the IFA shown in Figure 5.50.

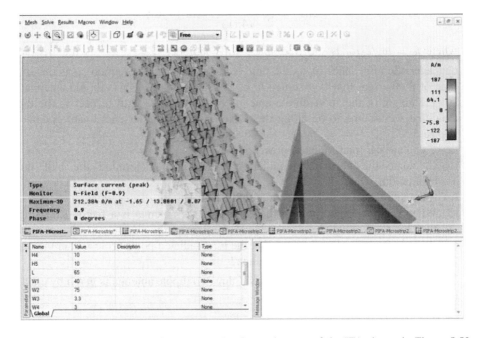

Figure 5.56 Detail of the surface current density at the rear of the IFA shown in Figure 5.50.

We see that the antenna structure is resonant (imaginary part of the input impedance equal to zero) at the desired frequency and that the real part of the input impedance is close to $50\,\Omega$. The surface current density of the structure at $900\,\mathrm{MHz}$ is shown in Figure 5.54.

The figure shows that current is flowing over the IFA giving rise to a dipole-like radiation pattern in the y-direction, see Figure 5.55.

The high surface current density on the microstrip trace shown in Figure 5.54 does not give rise to radiation, since this surface current density is compensated for by a surface current density on the microstrip ground plane that is concentrated at the strip position and directed in the opposite direction. This is shown in the detail of the surface current density at the rear of the antenna in Figure 5.56.

5.4 Problems

5.1 Plot, in a polar plot and on a linear scale, the normalized (power) radiation pattern of a small, z-directed dipole antenna as a function of φ for
   (a) $\vartheta = \frac{\pi}{4}$,
   (b) $\vartheta = \frac{\pi}{2}$.

5.2 The same question as in 5.1, but now plot in a rectangular plot and on a logarithmic scale.

5.3 The time-averaged radiated power density function of a short (Hertzian) dipole is given by

$$\mathbf{S}(\mathbf{r}) = \frac{1}{2}\frac{k_0^2 Z_0 \,(I_0 l)^2}{(4\pi r)^2}\sin^2(\vartheta)\hat{\mathbf{u}}_r,\tag{5.69}$$

where $k_0 = \omega\sqrt{\varepsilon_0\mu_0}$ is the free space wave number, $Z_0 = \sqrt{\frac{\mu_0}{\varepsilon_0}}$ is the free space characteristic impedance, $\hat{\mathbf{u}}_r$ is the unit vector in the radial direction, I_0 is the amplitude of the current that is assumed to be constant over the short dipole antenna and l is the length of the short dipole antenna. ϑ is the angle with respect to the dipole axis, that we assume to be along the z-axis in a rectangular coordinate system, see Figure 5.57.
   (a) Calculate the total radiated power P_t.
   (b) Calculate the normalized radiation pattern and determine the half power beamwidth.
   (c) Calculate the directivity function and the directivity of this antenna.
   (d) When the power accepted at the clamps of the antenna is a factor 1.001 higher than the total radiated power, what is the gain of the antenna?
   (e) What is, in general, the reason that the total radiated power is less than the power accepted at the clamps of an antenna?

5.4 The far magnetic field of an elementary z-directed dipole antenna is given by equation (5.11)

$$\mathbf{H} = \frac{jk_0 e^{-jk_0 r}}{4\pi r}I_0 l\,\sin\vartheta\,\hat{\mathbf{u}}_\varphi.\tag{5.70}$$

Explain why waves travel *away* from the source.

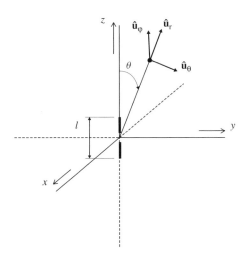

Figure 5.57 Short dipole antenna.

5.5 What current is needed to radiate 1 kW in total using a small dipole antenna having a length $l = 0.01\lambda_0$?

5.6 What current is needed to radiate 1 kW in total using a half-wave dipole antenna?

5.7 Draw, in the same rectangular plot, on a logarithmic scale, the normalized (power) radiation pattern of
   (a) an infinitesimal dipole antenna,
   (b) a half-wave dipole antenna,
   (c) a one-wavelength dipole antenna,
   (d) a $\frac{3}{2}$-wavelength antenna,
   for $\varphi = 0$ and $-\frac{\pi}{2} < \vartheta < \frac{\pi}{2}$ and determine from this graph the 3 dB beamwidths.

5.8 Plot the real and imaginary parts of a wire dipole antenna input impedance as a function of $k_0 l$ for $1.3 < k_0 l < 1.7$ and $a = 0.005\lambda_0$. Remember that l is the half-length of the dipole antenna.

5.9 At what length $2l$ is the dipole antenna from problem 5.8 resonant?

References

1. David M. Pozar, *Microwave Engineering, second edition*, John Wiley & Sons, New York, 1998.
2. Milton Abramowitz and Irene A. Stegun, *Handbook of Mathematical Functions*, Dover Publications, New York, 1972.
3. Robert S. Elliot, *Antenna Theory and Design, revised edition*, John Wiley & Sons, New York, 2003.
4. CST Computer Simulation Technology: http://www.cst.com
5. Constantine A. Balanis, *Antenna Theory, Analysis and Design, second edition*, John Wiley & Sons, New York, 1997.
6. Hans Schantz, *The Art and Science of Ultrawideband Antennas*, Artech House, Boston, 2005.
7. Hubregt J. Visser, *Approximate Antenna Analysis for CAD*, John Wiley & Sons, Chichester, UK, 2009.
8. E. Gueguen, F. Thudor and P. Chamberlin, 'A Low Cost UWB Printed Dipole Antenna with High Performances', *Proceedings of the IEEE International Conference on UWB*, pp. 89–92, 2005.

6

Loop Antennas

A loop antenna consists of a conductive wire, usually bent into a circular form, although other forms are possible. The small loop antenna characteristics are very similar to those of the small dipole antenna. The small loop antenna will be shown to be equivalent to an infinitesimal magnetic dipole. Loop antennas are denoted *small* or *large*, depending on the size of the circumference with respect to the wavelength. The derivation of the radiated fields of the loop antenna could be performed in a way similar to the procedure we used for calculating these fields for an infinitesimal z-directed current element: Obtain the vector potential for an infinitesimal, φ-directed current element and then generalize to a larger loop with an assumed current density. However, since we have already derived expressions for the far magnetic and electric fields for arbitrary current distributions, we will derive the far-field for a general loop, having a uniform current distribution, and then specify for small loops.

6.1 General Constant Current Loop

For the loop analysis we will consider a filamentary, circular wire loop of radius a in the xy-plane of a rectangular coordinate system, see Figure 6.1.

The radius can have any value, but we do assume the current to be uniform and in-phase over the loop since this simplifies the calculations considerably. Nevertheless, the loop with uniform current is valuable, for both small and large loops:

- For small loops, that is for circumferences smaller than a tenth of the free space wavelength, a constant current is a good approximation.
- For larger loops, a more realistic current density assessment would be in a cosh(l)-form (not a sinusoidal form [1]), where l is the distance parameter along the loop. A large loop supporting a uniform current may be realized though by subdividing the loop into smaller wire segments and feeding each segment (in amplitude and phase) in such a way that the amplitude and phase over the loop effectively remains constant [2].

Antenna Theory and Applications, First Edition. Hubregt J. Visser.
© 2012 John Wiley & Sons, Ltd. Published 2012 by John Wiley & Sons, Ltd.

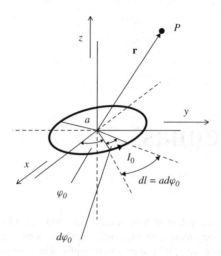

Figure 6.1 Wire loop of radius a, supporting a uniform, in-phase current.

6.1.1 *Radiation*

For the loop shown in Figure 6.1, we assume a (uniform) current density

$$\mathbf{J}_e\left(\mathbf{r}_0\right) = I_0 \delta \left(\sqrt{x_0^2 + y_0^2} - a\right) \delta\left(z_0\right) \hat{\mathbf{u}}_{\varphi_0}. \tag{6.1}$$

The far-fields for an arbitrary current density are calculated with equations (4.65) and (4.78)

$$\mathbf{H} = \frac{-jk_0 e^{-jk_0 r}}{4\pi r} \hat{\mathbf{u}}_r \times \iiint_{V_0} \mathbf{J}_e\left(\mathbf{r}_0\right) e^{jk_0 \hat{\mathbf{u}}_r \cdot \mathbf{r}_0} dV_0, \tag{6.2}$$

$$\mathbf{E} = Z_0 \mathbf{H} \times \hat{\mathbf{u}}_r, \tag{6.3}$$

where $Z_0 = \sqrt{\mu_0/\varepsilon_0}$ is the characteristic impedance of free space.

The integration over the source volume reduces to a contour integral over the loop:

$$\mathbf{H} = \frac{-jk_0 e^{-jk_0 r}}{4\pi r} I_0 \hat{\mathbf{u}}_r \times \int_l \hat{\mathbf{u}}_{\varphi_0} e^{jk_0 \hat{\mathbf{u}}_r \cdot \mathbf{r}_0} dl = \frac{-jk_0 e^{-jk_0 r}}{4\pi r} a I_0 \hat{\mathbf{u}}_r \times \int_{\varphi_0=0}^{2\pi} \hat{\mathbf{u}}_{\varphi_0} e^{jk_0 \hat{\mathbf{u}}_r \cdot \mathbf{r}_0} d\varphi_0, \tag{6.4}$$

where use has been made of, see Figure 6.1, $dl = a d\varphi_0$. To carry out the dot product in the exponent within the integral of equation (6.4), we first express $\hat{\mathbf{u}}_r$ and $\mathbf{r}_0$ in rectangular unit vectors.

$$\hat{\mathbf{u}}_r = \sin\vartheta\cos\varphi\hat{\mathbf{u}}_x + \sin\vartheta\sin\varphi\hat{\mathbf{u}}_y + \cos\vartheta\hat{\mathbf{u}}_z, \tag{6.5}$$

and, see Figure 6.1,

$$\mathbf{r}_0 = a\cos\varphi_0\hat{\mathbf{u}}_x + a\sin\varphi_0\hat{\mathbf{u}}_y, \tag{6.6}$$

so that

$$\hat{\mathbf{u}}_r \cdot \mathbf{r}_0 = a \cos \varphi \cos \varphi_0 \sin \vartheta + a \sin \varphi \sin \varphi_0 \sin \vartheta = a \cos(\varphi - \varphi_0) \sin \vartheta. \tag{6.7}$$

Substitution of equation (6.7) in equation (6.4) results in

$$\mathbf{H} = \frac{jk_0 e^{-jk_0 r}}{4\pi r} a I_0 \hat{\mathbf{u}}_r \times \int_{\varphi_0=0}^{2\pi} \hat{\mathbf{u}}_{\varphi_0} e^{jk_0 a \cos(\varphi - \varphi_0) \sin \vartheta} d\varphi_0. \tag{6.8}$$

Since the structure being analyzed is rotationally symmetric around the z-axis in Figure 6.1 and thus φ-independent, we may substitute–without loss of generality – $\varphi = 0$ and obtain

$$\mathbf{H} = \frac{jk_0 e^{-jk_0 r}}{4\pi r} a I_0 \hat{\mathbf{u}}_r \times \int_{\varphi_0=0}^{2\pi} \hat{\mathbf{u}}_{\varphi_0} e^{jk_0 a \cos \varphi_0 \sin \vartheta} d\varphi_0. \tag{6.9}$$

With the relation between cylindrical and rectangular coordinates given by [3]

$$\hat{\mathbf{u}}_\rho = \cos \varphi \hat{\mathbf{u}}_x + \sin \varphi \hat{\mathbf{u}}_y, \tag{6.10}$$

$$\hat{\mathbf{u}}_\varphi = -\sin \varphi \hat{\mathbf{u}}_x + \cos \varphi \hat{\mathbf{u}}_y, \tag{6.11}$$

$$\hat{\mathbf{u}}_z = \hat{\mathbf{u}}_z, \tag{6.12}$$

we may express $\hat{\mathbf{u}}_{\varphi_0}$ in equation (6.9) as $\hat{\mathbf{u}}_{\varphi_0} = -\sin \varphi_0 \hat{\mathbf{u}}_x + \cos \varphi_0 \hat{\mathbf{u}}_y$, so that

$$\mathbf{H} = \frac{-jk_0 e^{-jk_0 r}}{4\pi r} a I_0 \hat{\mathbf{u}}_r \times \left\{ -\hat{\mathbf{u}}_x \int_0^{2\pi} \sin \varphi_0 e^{jk_0 a \cos \varphi_0 \sin \vartheta} d\varphi_0 \right.$$
$$\left. + \hat{\mathbf{u}}_y \int_0^{2\pi} \cos \varphi_0 e^{jk_0 a \cos \varphi_0 \sin \vartheta} d\varphi_0 \right\}. \tag{6.13}$$

The first integral on the right-hand side of equation (6.13) equals zero since the kernel of the integral is an odd function of φ_0.[1] The kernel of the second integral is an even function of φ_0 so that we find for the far magnetic field

$$\mathbf{H} = 2 \frac{-jk_0 e^{-jk_0 r}}{4\pi r} a I_0 \hat{\mathbf{u}}_r \times \hat{\mathbf{u}}_y \int_{\varphi_0=0}^{\pi} \cos \varphi_0 e^{jk_0 a \cos \varphi_0 \sin \vartheta} d\varphi_0$$

$$= 2 \frac{-jk_0 e^{-jk_0 r}}{4\pi r} I_0 \hat{\mathbf{u}}_y \int_{\varphi_0-\frac{\pi}{2}=0}^{\pi} \cos\left(\varphi_0 - \frac{\pi}{2}\right) e^{jk_0 a \cos(\varphi_0 - \frac{\pi}{2}) \sin \vartheta} d\left(\varphi_0 - \frac{\pi}{2}\right)$$

$$= -2 \frac{-jk_0 e^{-jk_0 r}}{4\pi r} a I_0 \hat{\mathbf{u}}_r \times \hat{\mathbf{u}}_y \int_{\varphi_0=-\frac{\pi}{2}}^{\frac{\pi}{2}} \sin \varphi_0 e^{-jk_0 a \sin \varphi_0 \sin \vartheta} d\varphi_0$$

$$= -2 \frac{-jk_0 e^{-jk_0 r}}{4\pi r} a I_0 \hat{\mathbf{u}}_r \times \hat{\mathbf{u}}_y \left\{ \int_{\varphi_0=-\frac{\pi}{2}}^{0} \sin \varphi_0 e^{-jk_0 a \sin \varphi_0 \sin \vartheta} d\varphi_0 \right.$$

---

[1] a function $f(x)$ is odd if $-f(x) = f(-x)$ or $f(x) + f(-x) = 0$. The function is even if $f(x) = f(-x)$ or $f(x) + f(-x) = 2f(x)$.

$$+ \int_{\varphi_0=0}^{\frac{\pi}{2}} \sin \varphi_0 e^{-jk_0 a \sin \varphi_0 \sin \vartheta} d\varphi_0 \Bigg\}$$

$$= -2 \frac{-jk_0 e^{-jk_0 r}}{4\pi r} a I_0 \hat{\mathbf{u}}_r \times \hat{\mathbf{u}}_y \left\{ -\int_{\varphi_0=0}^{-\frac{\pi}{2}} \sin \varphi_0 e^{-jk_0 a \sin \varphi_0 \sin \vartheta} d\varphi_0 \right.$$

$$\left. + \int_{\varphi_0=0}^{\frac{\pi}{2}} \sin \varphi_0 e^{-jk_0 a \sin \varphi_0 \sin \vartheta} d\varphi_0 \right\}$$

$$= -2 \frac{-jk_0 e^{-jk_0 r}}{4\pi r} a I_0 \hat{\mathbf{u}}_r \times \hat{\mathbf{u}}_y \left\{ -\int_{-\varphi_0=0}^{\frac{\pi}{2}} \sin(-\varphi_0) e^{+jk_0 a \sin(-\varphi_0) \sin \vartheta} d(-\varphi_0) \right.$$

$$\left. + \int_{\varphi_0=0}^{\frac{\pi}{2}} \sin \varphi_0 e^{-jk_0 a \sin \varphi_0 \sin \vartheta} d\varphi_0 \right\}$$

$$= -2 \frac{jk_0 e^{-jk_0 r}}{4\pi r} a I_0 \hat{\mathbf{u}}_r \times \hat{\mathbf{u}}_y \int_0^{\frac{\pi}{2}} \sin \varphi_0 \left(e^{+jk_0 a \sin \varphi_0 \sin \vartheta} - e^{-jk_0 a \sin \varphi_0 \sin \vartheta} \right) d\varphi_0$$

$$= 4 \frac{k_0 e^{-jk_0 r}}{4\pi r} a I_0 \hat{\mathbf{u}}_r \times \hat{\mathbf{u}}_y \int_0^{\frac{\pi}{2}} \sin \varphi_0 \sin(k_0 a \sin \varphi_0 \sin \vartheta) d\varphi_0. \qquad (6.14)$$

In equation (6.14) we may recognize a Bessel function of the first kind and order [1, 4]

$$J_1(x) = \frac{2}{\pi} \int_0^{\frac{\pi}{2}} \sin \vartheta \sin(x \sin \vartheta) d\vartheta. \qquad (6.15)$$

Figure 6.2 shows $J_1(x)$ vs x for $0 \le x \le 50$.

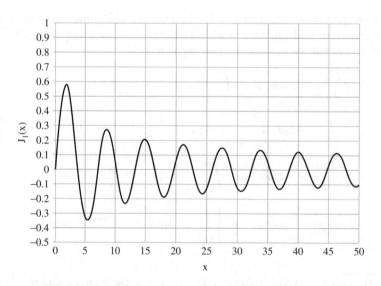

Figure 6.2 Bessel function of the first kind, order 1 and argument x for $0 \le x \le 50$.

Equation (6.14), written in terms of this Bessel function now becomes

$$\mathbf{H} = \hat{\mathbf{u}}_r \times \hat{\mathbf{u}}_y \frac{4k_0 e^{-jk_0 r}}{4\pi r} a I_0 \frac{\pi}{2} J_1\left(k_0 a \sin \vartheta\right) = \hat{\mathbf{u}}_r \times \hat{\mathbf{u}}_y \frac{k_0 a I_0 e^{-jk_0 r}}{2r} J_1\left(k_0 a \sin \vartheta\right). \quad (6.16)$$

Earlier we made the choice to work in the plane $\varphi = 0$. For $\varphi = 0$, see Figure 6.1, $\hat{\mathbf{u}}_y = \hat{\mathbf{u}}_\varphi$. With $\hat{\mathbf{u}}_r \times \hat{\mathbf{u}}_\varphi = -\hat{\mathbf{u}}_\vartheta$, we finally find for the far magnetic field

$$\mathbf{H} = -\hat{\mathbf{u}}_\vartheta \frac{k_0 a I_0 e^{-jk_0 r}}{2r} J_1\left(k_0 a \sin \vartheta\right). \quad (6.17)$$

The far electric field we find by substitution of equation (6.17) in (6.3). Also using $-\hat{\mathbf{u}}_\vartheta \times \hat{\mathbf{u}}_r = \hat{\mathbf{u}}_\varphi$ results in

$$\mathbf{E} = \hat{\mathbf{u}}_\varphi \frac{k_0 a Z_0 I_0 e^{-jk_0 r}}{2r} J_1\left(k_0 a \sin \vartheta\right). \quad (6.18)$$

The normalized radiation pattern follows from equation (6.18)

$$F(\vartheta) = \frac{P(\vartheta)}{P(\vartheta)_{max}} = \frac{r^2 |\mathbf{E}|/(2Z_0)}{r^2 |\mathbf{E}|_{max}/(2Z_0)} = \frac{J_1^2\left(k_0 a \sin \vartheta\right)}{\left[J_1^2\left(k_0 a \sin \vartheta\right)\right]_{max}}. \quad (6.19)$$

Figure 6.3 shows the radiation patterns for $a = \frac{\lambda_0}{20}$, for $a = \frac{\lambda_0}{2}$ and for $a = 2\lambda_0$.

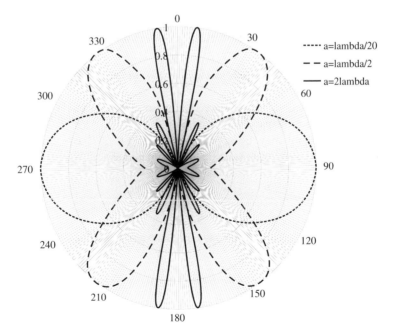

Figure 6.3 Normalized (power) radiation patterns for a horizontal, constant current supporting, loop antenna. Dotted line: $a = \frac{\lambda_0}{20}$. Dashed line: $a = \frac{\lambda_0}{2}$. Solid line: $a = 2\lambda_0$.

The figure shows that for a very small loop, the radiation pattern looks very similar to that of a vertically directed elementary dipole. We will discuss the similarities between a small loop and a small dipole in more detail in a subsequent section.

We also see that for larger loops – for which we have to take measures to keep the current density uniform and in-phase – we get lobes in the forward direction and a sharp null perpendicular to the loop. This feature makes large loop antennas attractive for direction finding purposes. It is easier to detect a null than a maximum.

6.1.2 Input Impedance

For calculating the radiation resistance, we need to know the total radiated power P_t. By using equations (2.8) and (6.18) we calculate P_t as

$$P_t = \int_{\varphi=0}^{2\pi} \int_{\vartheta=0}^{\pi} P(\vartheta) \sin \vartheta \, d\vartheta \, d\varphi$$

$$= \int_{\varphi=0}^{2\pi} \int_{\vartheta=0}^{\pi} \frac{r^2 |\mathbf{E}|^2}{2Z_0} \sin \vartheta \, d\vartheta \, d\varphi$$

$$= \frac{\pi (k_0 a I_0)^2 Z_0}{4} \int_0^{\pi} J_1^2 (k_0 a \sin \vartheta) \, d\vartheta$$

$$= 30\pi^2 (k_0 a I_0)^2 \int_0^{\pi} J_1^2 (k_0 a \sin \vartheta) \, d\vartheta. \tag{6.20}$$

The integral over the square of the Bessel function of order 1 may be transformed into an integral over a Bessel function of order 2. We then find [5, 6]

$$P_t = 30\pi^2 k_0 a I_0^2 \int_0^{2k_0 a} J_2(y) dy. \tag{6.21}$$

Note that the upper integration boundary in equation (6.21) is twice the circumference of the loop expressed in free-space wavelengths.

For $k_0 a \geq 5$ the integral may be approximated as [5] $\int_0^{2k_0 a} J_2(y) dy \approx 1$. The total radiated power for large loops may thus be approximated as

$$P_t = 30\pi^2 k_0 a I_0^2 = \frac{1}{2} I_0^2 R_A, \tag{6.22}$$

where R_A is the radiation resistance.

For large loops, supporting a uniform, in-phase current density, the radiation resistance is thus found as

$$R_A = 60\pi^2 k_0 a. \tag{6.23}$$

So, for a loop with a circumference of five wavelengths ($2\pi a = 5\lambda_0$), the radiation resistance is thus $R_A = 60\pi^2 \cdot 5 = 2961 \, \Omega$.

The loop antenna may be envisaged as a series circuit consisting of a resistor (R_A) and an inductor. The inductor value is given by [7]

$$L = \mu_0 a \left[\ln \left(\frac{8a}{b} \right) - 2 \right], \tag{6.24}$$

where a is the radius of the loop and b is the radius of the wire.

6.1.3   Small Loop Antenna

The far magnetic and electric fields for a small loop antenna $(2\pi a \leq \lambda_0/10)$ are obtained from equations (6.17) and (6.18), respectively, by using a small-argument approximation for the Bessel function: $J_1(z) \approx \frac{z}{2}$ for small z.

Thus, we obtain for a small loop antenna:

$$\mathbf{H} = -\frac{k_0^2 \left(\pi a^2 I_0 \right) e^{-jk_0 r}}{4\pi r} \sin \vartheta \, \hat{\mathbf{u}}_\vartheta, \tag{6.25}$$

$$\mathbf{E} = \frac{k_0^2 Z_0 \left(\pi a^2 I_0 \right) e^{-jk_0 r}}{4\pi r} \sin \vartheta \, \hat{\mathbf{u}}_\varphi. \tag{6.26}$$

For calculating the normalized (power) radiation pattern we need to calculate the radiated power per unit of solid angle. By using equation (2.26) we find

$$P(\vartheta) = \frac{r^2}{2Z_0} |\mathbf{E}|^2 = \frac{k_0^4 Z_0 a^4 I_0^2}{32} \sin^2 \vartheta, \tag{6.27}$$

and for the normalized radiation pattern

$$F(\vartheta) = \sin^2 \vartheta. \tag{6.28}$$

Based on the radiation pattern of a loop having radius $a = \lambda_0/20$, shown in Figure 6.3, we have already concluded that the radiation pattern of a small loop antenna is very similar to that of a small dipole antenna. Now, based on equation (6.28), we may conclude that the radiation pattern of a horizontal, small loop antenna is *identical* to that of a small, vertical dipole antenna.

For calculating the radiation resistance, we first have to calculate the total radiated power

$$\begin{aligned} P_t &= \int_{\varphi=0}^{2\pi} \int_{\vartheta=0}^{\pi} \frac{k_0^4 Z_0 a^4 I_0^2}{32} \sin^3 \vartheta \, d\varphi \\ &= \frac{\pi k_0^4 Z_0 a^4 I_0^2}{16} \int_0^{\pi} \sin^3 \vartheta \, d\vartheta \\ &= \frac{\pi k_0^4 Z_0 a^4 I_0^2}{12}, \end{aligned} \tag{6.29}$$

where use has been made of $\int_0^{\pi} \sin^3 \vartheta \, d\vartheta = \frac{4}{3}$.

The radiation resistance R_A is then found as

$$R_A = \frac{P_t}{\frac{1}{2}I_0^2} = \frac{\pi k_0^4 Z_0 a^4}{6} = 320\pi^4 \left[\pi \left(\frac{a}{\lambda_0} \right)^2 \right]^2 . \tag{6.30}$$

For a loop having radius $a = \lambda_0/20$, the radiation resistance is found to be $R_A = 1.92\,\Omega$. The reactance is found with equation (6.24).

The calculation of the directivity is now straight forward and left as an exercise to the reader.

6.1.4 Comparison of Short Dipole and Small Loop Antenna

We will now compare a small z-directed dipole antenna with a small loop antenna in the xy-plane, see Figure 6.4.

The far-field components of the elementary dipole are given by

$$H_{\varphi_d} = \frac{jk_0 p e^{-jk_0 r}}{4\pi r} \sin\vartheta, \tag{6.31}$$

$$E_{\vartheta_d} = \frac{jk_0 Z_0 p e^{-jk_0 r}}{4\pi r} \sin\vartheta, \tag{6.32}$$

where $p = I_0 l$ is known as the *transmitter moment* of the dipole antenna.

The far-field components of the elementary loop antenna are given by

$$H_{\vartheta_l} = -\frac{k_0^2 m e^{-jk_0 r}}{4\pi r} \sin\vartheta, \tag{6.33}$$

$$E_{\varphi_l} = \frac{k_0^2 Z_0 m e^{-jk_0 r}}{4\pi r} \sin\vartheta, \tag{6.34}$$

where $m = I_0 \pi a^2$ is the transmitter moment of the loop antenna.

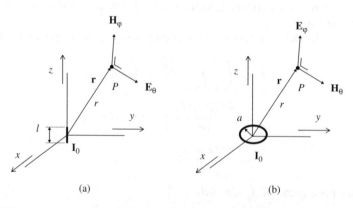

(a)   (b)

Figure 6.4 Elementary antennas. (a) Elementary, vertical dipole antenna. (b) Elementary, horizontal loop antenna.

From equations (6.31)–(6.34) we find that if $p = -mk_0$:

$$H_{\varphi_d} = jH_{\vartheta_l}, \qquad (6.35)$$

$$E_{\vartheta_d} = -jE_{\varphi_l}. \qquad (6.36)$$

The elementary loop antenna is thus the dual of the elementary dipole antenna.

6.2 Printed Loop Antenna

In this section we will design a rectangular strip loop antenna on a PCB for a remote keyless system. The operational frequency will be 433.92 MHz, a frequency dedicated in Europe for these systems. We will start with what we have learnt in this chapter to come up with an initial design that we will fine-tune, using the CST-MWS full wave analysis software.

6.2.1 Application of Theory

For a remote keyless system, the transmitter must be very small and fit in a user's hand. The wavelength to be used is 69.14 cm, so we need to employ an antenna that is much smaller than the wavelength. The small loop antenna (circumference smaller than a tenth of a wavelength) is providing a solution to this design problem.

The radiation resistance of a small loop antenna will be very low as we have seen in section 6.1.3. The value may in fact be lower than the loss resistance of the metal conductor, thus leading to a low radiation efficiency. Nevertheless, we will use a small loop antenna due to the size benefit and the closest to omnidirectional radiation pattern it provides. As we will see, we will be able to impedance match the antenna to 50 Ω, using only two lumped element capacitors.

Since for a small loop antenna we may consider the current flowing through it to be constant, and since the dimensions are much smaller than a wavelength, the actual shape of the loop is of little importance. For analysis and manufacturing ease we therefore choose the loop to be square, see Figure 6.5.

The radiation resistance R_A of this loop antenna follows from equation (6.30) upon substituting $\left(\frac{P}{4}\right)^2$ for πa^2 (the physical area occupied by the loop antenna).

$$R_A = \frac{5}{4}\left(\frac{\pi P}{\lambda}\right)^4, \qquad (6.37)$$

provided that the perimeter is less than a tenth of a wavelength, $P < \frac{\lambda}{10}$.

The loss resistance R_l of the metal strip having width W and a conductivity σ at a frequency f is given by [8]

$$R_l = \frac{P}{2W}\sqrt{\frac{\pi f \mu_0}{\sigma}} = \frac{P\pi}{2W}\sqrt{\frac{120}{\lambda\sigma}}. \qquad (6.38)$$

An improved loss resistance equation, accounting for the edge effect of the strip [9], is given by $R_l' = \kappa R_l$, where $1 < \kappa < 2$.

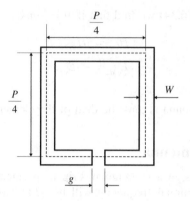

Figure 6.5 Square strip loop antenna having side $\frac{P}{4}$, strip width W and excitation gap width g.

The square loop inductance L is given by [10]

$$L = \frac{\mu_0 P}{2\pi} \ln \left(\frac{8 \left(\frac{P}{4} \right)^2}{P W} \right) = \frac{\mu_0 P}{2\pi} \ln \left(\frac{P}{2W} \right). \qquad (6.39)$$

We now have the material necessary to calculate the capacitor values in a two-capacitor impedance matching network.

6.2.1.1 Loop Antenna Matching

The loop antenna may be seen as a series electrical circuit consisting of an inductance L and a resistor R, see Figure 6.6a. The resistor R comprises the radiation loss R_A and the conductor loss R_l.

With the aid of two capacitors we may impedance match this antenna to a desired impedance value, in the process boosting the resistance value [11]. Before we describe this matching process we need to be able to transform a series impedance into a parallel impedance.

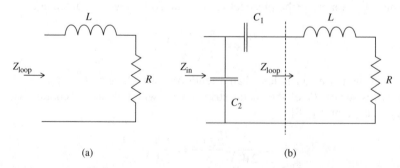

(a)   (b)

Figure 6.6 Equivalent electric circuit of a small loop antenna (a) and the impedance matching of this antenna (b).

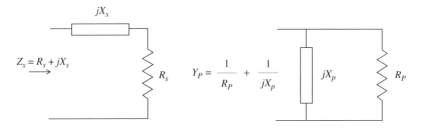

Figure 6.7   Series impedance and equivalent parallel impedance.

Figure 6.7 shows a general series impedance and the equivalent parallel impedance. R_p and X_p can be expressed in terms of R_s and X_s by calculating the admittances of both circuits:

$$Y_s = \frac{1}{R_s + jX_s} = Y_p = \frac{1}{R_p} + \frac{1}{jX_p}. \tag{6.40}$$

Multiplying the nominator and denominator of Y_s by $(R_s - jX_s)$ results in

$$\frac{R_s - jX_s}{R_s^2 + X_s^2} = \frac{1}{R_p} - j\frac{1}{X_p}, \tag{6.41}$$

so that

$$R_p = \frac{R_s^2 + X_s^2}{R_s}, \tag{6.42}$$

$$X_p = \frac{R_s^2 + X_s^2}{X_s}. \tag{6.43}$$

By introducing the series circuit quality factor

$$Q_s = \frac{X_s}{R_s}, \tag{6.44}$$

we obtain

$$R_p = R_s \left(1 + Q_s^2\right), \tag{6.45}$$

$$X_p = X_s \left(\frac{1 + Q_s^2}{Q_s^2}\right). \tag{6.46}$$

We will now perform the impedance matching in two steps. First, we will introduce the series capacitor C_1 only, see Figure 6.6, and find the value for C_1 that boosts the real part of the loop impedance to the desired value. Then we will add the second, parallel capacitor C_2 to tune out the remaining imaginary part of the circuit formed by the small loop antenna and the series capacitor C_1.

We add series capacitor C_1 to the loop antenna, getting the equivalent series circuit shown in Figure 6.8(a) and then transform this circuit into an equivalent parallel circuit as shown in Figure 6.8(b).

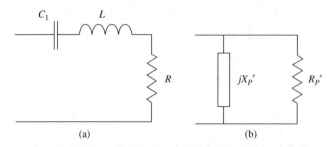

(a)   (b)

Figure 6.8   Series circuit representation of a small loop antenna with series capacitor C_1 (a) and the equivalent parallel circuit (b).

The input impedance of the circuit is

$$Z_{in} = R_{in} + jX_{in},$$ (6.47)

where, with the aid of equation (6.45),

$$R_{in} = R\left(1 + Q_s^2\right).$$ (6.48)

Since R_{in} and R are known (R_{in} is the desired real part of the input impedance), we find for Q_s

$$Q_s = \sqrt{\frac{R_{in}}{R} - 1}.$$ (6.49)

The series quality factor was defined as the ratio of the imaginary and the real part of the series impedance, see equation (6.44), so

$$Q_s = \frac{X_s}{R_s} = \frac{\omega L - \frac{1}{\omega C_1}}{R}.$$ (6.50)

From equations (6.49) and (6.50) we find the value for C_1 that will boost the real part of the loop impedance to the desired value

$$C_1 = \frac{1}{\omega\left(\omega L - R\sqrt{\frac{R_{in}}{R} - 1}\right)}.$$ (6.51)

With C_1 added in series with the small loop antenna, the parallel reactance X_p' (see Figure 6.8) is found with the aid of equations (6.44), (6.46) and (6.50) as

$$X_p' = Q_s R_s \frac{1 + Q_s^2}{Q_s^2} = R\frac{1 + Q_s^2}{Q_s}$$

$$= R\frac{1 + \frac{R_{in}}{R} - 1}{\sqrt{\frac{R_{in}}{R} - 1}} = \frac{R_{in}}{\sqrt{\frac{R_{in}}{R} - 1}}.$$ (6.52)

Since $R_{in} > R$, then $X_p' > 0$ and it can indeed be tuned out with a parallel capacitor C_2. The value of C_2, see Figure 6.6, follows from

$$1 - \omega C_2 X_p' = 0, \tag{6.53}$$

so that

$$C_2 = \frac{1}{\omega X_p'} = \frac{\sqrt{\frac{R_{in}}{R} - 1}}{\omega R_{in}}. \tag{6.54}$$

6.2.2 Design of a Printed Loop Antenna

With the analytical equations for the small, square, strip loop antenna and the two-capacitor matching network we can design the antenna for 433.92 MHz, impedance matched for a balanced 50 Ω front-end.

We start with a square loop antenna having $\frac{P}{4} = 17$ mm, see Figure 6.5, so that the perimeter of the loop is smaller than a tenth of a wavelength at the operational frequency. We choose the strip width to be $W = 1.5$ mm, see Figure 6.5, and made of copper ($\sigma = 5.8 \cdot 10^7$ Sm^{-1}) of 70 μm thickness. For $W = 1.5$ mm, κ equals 1.75.[2]

With the aid of equations (6.37), (6.38) and $R_l' = \kappa R_l$, we find that

$$R_A = 0.011\ \Omega, \tag{6.55}$$

$$R_l' = 0.21\ \Omega, \tag{6.56}$$

and for the radiation efficiency

$$\eta = \frac{R_A}{R_A + R_l'} = 0.050. \tag{6.57}$$

So, the real part of the input impedance of the loop equals

$$R_{loop} = R_A + R_l' = 0.22\ \Omega, \tag{6.58}$$

and the imaginary part is calculated with equation (6.39) as

$$X_{loop} = \omega L = 115.72, \tag{6.59}$$

at 433.92 MHz.

Next, we will analyze this loop antenna in CST-MWS for verification of the analytical equations. The loop is drawn as shown in Figure 6.9. The loop is excited with a discrete port over the (1 mm wide) gap.

The loop is analyzed in the frequency domain, and in the menu *results* (top bar) the option *S Parameter Calculations* is chosen to calculate the Z-matrix. After finishing these calculations, from the menu on the left we then choose *1D Results*, *Z Matrix*, *Real Part* and *Imaginary Part*. Using the *Plot Properties* – through a right mouse click on the

---

[2] This value is obtained through an intensive comparison of loops of different perimeter lengths, calculated analytically and analyzed full-wave employing CST-MWS.

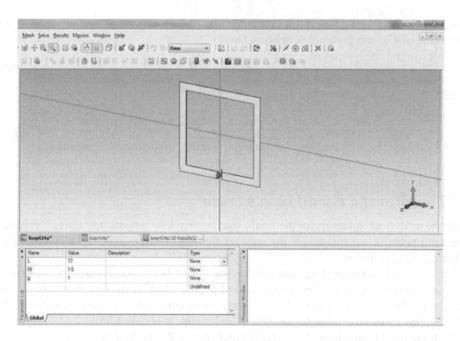

Figure 6.9 Layout of a square strip loop antenna, excited with a discrete port over a gap.

plot – to zoom in on the results and the *Show Measure Lines* (right mouse click again) to show specific values, we get the results shown in Figures 6.10 and 6.11 for the real and imaginary parts of the loop input impedance as a function of frequency, respectively.

From the figures we see that, at the operational frequency, the real part of the total loop impedance equals 0.23 Ω and the imaginary part equals 112.5 Ω. Taking these values as reference, the analytical calculation of the real part of the input impedance (0.22 Ω) has a relative error of 4.3%. The analytical calculation of the imaginary part (115.7 Ω) results in a relative error of 2.8%. Both errors are considered to be small enough for initial design purposes.

Finally, in Figure 6.12 we show the calculated far-field radiation (selected by adding a field monitor in the menu on the left before analysis).

The figure shows a directivity of 2.09 dBi ($= 10^{2.09/10} = 1.6$), which is slightly larger than the expected value ($1.5 = 10 \log(1.5)$ dBi $= 1.8$ dBi) and must be attributed to the fact that we are at the perimeter boundary of what may be considered a *small* loop antenna. The radiation efficiency is -13.05 dB ($= 10^{-13.05/10} = 0.050$) and agrees very well with the analytical calculation result, equation (6.57).

With the analytical equations thus being verified, we may now proceed by impedance matching this antenna to $R_{in} = 50\,\Omega$.

Using equations (6.51) and (6.54), for C_1 and C_2, respectively leads to

$$C_1 = 3.36 \cdot 10^{-12}\,\mathrm{F} = 3.36\,\mathrm{pF}, \tag{6.60}$$

$$C_2 = 1.0791 \cdot 10^{-10}\,\mathrm{F} = 107.91\,\mathrm{pF}. \tag{6.61}$$

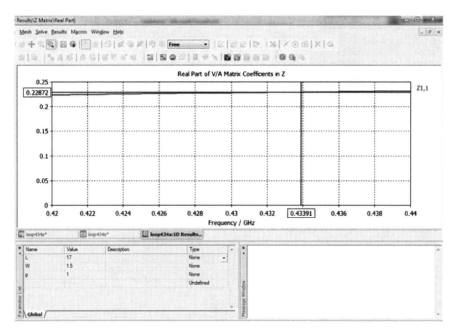

Figure 6.10 Real part of the input impedance as a function of frequency of the square strip loop antenna shown in Figure 6.9.

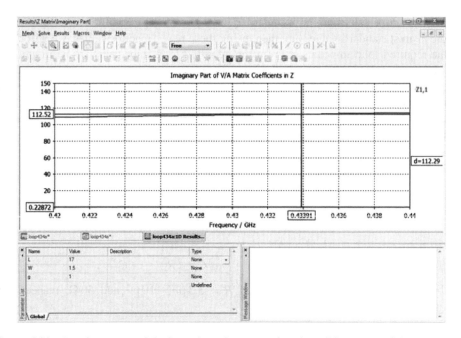

Figure 6.11 Imaginary part of the input impedance as a function of frequency of the square strip loop antenna shown in Figure 6.9.

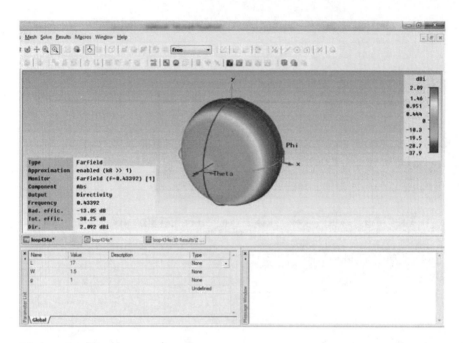

Figure 6.12   Radiation pattern of the square strip loop antenna shown in Figure 6.9.

These capacitor values are not standard values. Therefore, we will change these values into the nearest standard values and then, using these new capacitor values, we will calculate the required loop impedance. From this required loop impedance we will determine the side, $\frac{P}{4}$, and the width, W, of the loop. In this determination we will introduce an approximation. To check the influence of this approximation, we will afterwards calculate the input impedance of the matched loop.

The nearest standard capacitor values for C_1 and C_2 are 3.3 pF and 110 pF, respectively. Then, from Figure 6.6(b), with $Z_{in} = R_{in}$, we calculate R and L as

$$R = \frac{R_{in}\left(1 + \frac{C_2}{C_1}\right)}{1 + \omega^2 C_2^2 R_{in}^2} - \frac{\omega^2 C_1 C_2 R_{in}}{\omega^4 C_1^2 C_2^2 R_{in}^2 + \omega^2 C_1^2}, \tag{6.62}$$

and

$$L = \frac{C_2 R_{in}^2 \left(1 + \frac{C_2}{C_1}\right)}{1 + \omega^2 C_2^2 R_{in}^2} + \frac{C_1}{\omega^4 C_1^2 C_2^2 R_{in}^2 + \omega^2 C_1^2}. \tag{6.63}$$

Substituting the values for C_1, C_2 and R_{in}: 3.3 pF, 110 pF and 50 Ω, respectively, leads to

$$R = 0.22\,\Omega, \tag{6.64}$$

$$L = 41.98\,nH; \omega L = 114.45\,\Omega. \tag{6.65}$$

Now, we approximate the loop resistance as consisting solely of the loss contribution. Then, from equation (6.38) we find

$$\frac{P}{2W} = \alpha = \frac{R}{\pi}\sqrt{\frac{\lambda\sigma}{120}}, \tag{6.66}$$

and from equation (6.39) we find

$$P = \frac{2\pi L}{\mu_0 \ln(\alpha)}, \tag{6.67}$$

so that from α and P we obtain

$$W = \frac{P}{2\alpha}. \tag{6.68}$$

Substituting the values for R, L, λ and σ ($\sigma = 5.8 \cdot 10^7\,\mathrm{Sm}^{-1}$) in the above equations gives

$$\frac{P}{4} = 14.16\,\mathrm{mm}, \tag{6.69}$$

$$W = 0.70\,\mathrm{mm}. \tag{6.70}$$

To check the validity of our approximation for the loop resistance we will now calculate both the radiation resistance and the loss resistance using the found values for P and W, using equations (6.37) and (6.38). The input impedance of the capacitor-matched circuit is then calculated as

$$Z_{in} = \frac{\frac{1}{j\omega C_2}\left(\frac{1}{j\omega C_1} + j\omega L + R_A + R_l\right)}{\frac{1}{j\omega C_2} + \frac{1}{j\omega C_1} + j\omega L + R_A + R_l}. \tag{6.71}$$

Substituting all the found values results in $Z_{in} = (48.24 - j1.01)\,\Omega$, a value considered close enough to the desired $50\,\Omega$ input impedance.

Next, the square loop with $\frac{P}{4} = 14.16\,\mathrm{mm}$ and $W = 0.70\,\mathrm{mm}$ is analyzed in CST Microwave Studio®. The real part of the input impedance as a function of frequency is shown in Figure 6.13 and the imaginary part of the input impedance as a function of frequency is shown in Figure 6.14.

The figures reveal a discrepancy between the approximate model and the full-wave model. The full wave analysis results are $R = 0.36\,\Omega$ and that $\omega L = 116\,\Omega$.

We will use the full wave analysis software to tune the loop for getting the desired R and ωL values. To increase the ωL value, we need to enlarge $\frac{P}{4}$. To compensate for the fact that this will also increase the R value and for obtaining a lower R value, we will also widen the strip width W.

After a few iterations, we find for $\frac{P}{4} = 15.5\,\mathrm{mm}$ and $W = 1.5\,\mathrm{mm}$ that $R = 0.23\,\Omega$ and $\omega L = 114\,\Omega$, which we consider as acceptable results.

However, this cannot be the final design. In a practical application, the loop needs to be attached to some form of carrier. In this example we choose to use a low-loss (PCB) substrate having $\varepsilon_r = 3.55$ and $\tan\delta = 0.0021$ with a thickness of $1.524\,\mathrm{mm}$. We take

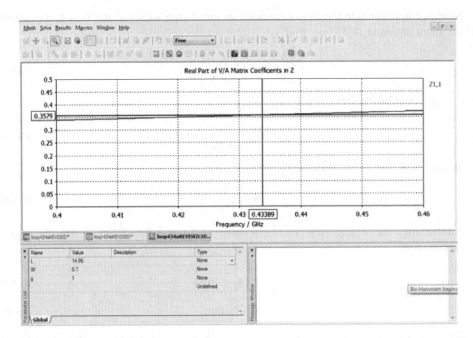

Figure 6.13 Real part of the input impedance as a function of frequency for a square strip loop antenna with $\frac{P}{4} = 14.16$ mm and $W = 0.70$ mm.

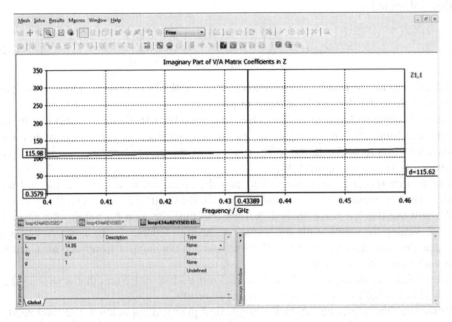

Figure 6.14 Imaginary part of the input impedance as a function of frequency for a square strip loop antenna with $\frac{P}{4} = 14.16$ mm and $W = 0.70$ mm.

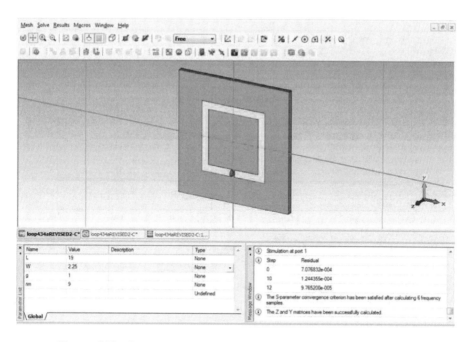

Figure 6.15 Square loop printed strip antenna on a low-loss substrate.

the $\frac{P}{4}$ and W values just found for a free-standing square strip loop as initial values and tune these values to realize the desired input impedance. The final structure is shown in Figure 6.15. The dimensions of the square strip (of thickness 70 μm) are: $\frac{P}{4} = 16.75$ mm and $W = 2.25$ mm. The rim of the substrate is 9 mm wide. The real and imaginary part of the input impedance as a function of frequency are shown in Figures 6.16 and 6.17, respectively.

As the last two figures show, at 433.92 MHz we obtain: $R = 0.235\,\Omega$ and $\omega L = 114.58\,\Omega$. Upon substituting these values in equation (6.71), together with $C_1 = 3.3$ pF and $C_2 = 110$ pF, we get for the matched loop input impedance $Z_{in} = (40.20 - j20.25)\,\Omega$. This impedance gives rise to a reflection coefficient amplitude, when connected to a 50 Ω system, upon substitution in equation (2.22) of $|\Gamma| = 0.24 = -12.4$ dB, which we consider as acceptable.

6.3 Problems

6.1 What current is needed to radiate 1 kW in total using a large, uniform current, loop antenna, having a circumference $2\pi a = 5\lambda_0$?

6.2 What current is needed to radiate 1 kW in total using a small loop antenna, having a circumference $2\pi a = \frac{\lambda_0}{20}$?

6.3 Calculate the directivity of a small loop antenna.

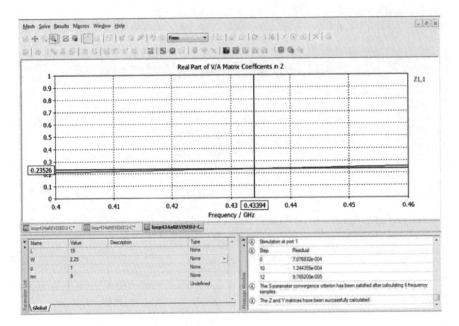

Figure 6.16 Real part of the input impedance as a function of frequency for the antenna shown in Figure 6.15.

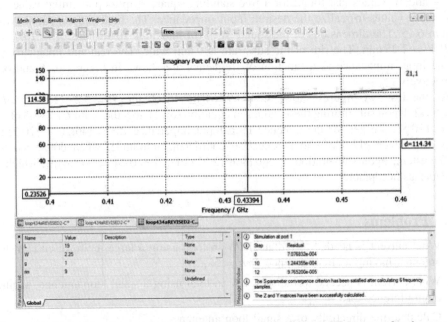

Figure 6.17 Imaginary part of the input impedance as a function of frequency for the antenna shown in Figure 6.15.

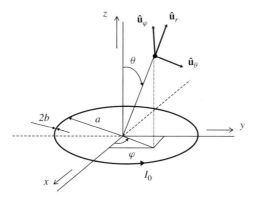

Figure 6.18   Small loop antenna.

6.4 A small wire loop antenna having radius a (where $2\pi a < \frac{\lambda_0}{10}$) has a far electric field given by

$$\mathbf{E} = \frac{k_0^2 Z_0 \left(\pi a^2 I_0\right) e^{-jk_0 r}}{4\pi r} \sin \vartheta \, \hat{\mathbf{u}}_\varphi, \tag{6.72}$$

where $k_0 = \omega \sqrt{\varepsilon_0 \mu_0}$ is the free space wave number, $Z_0 = \sqrt{\frac{\mu_0}{\varepsilon_0}}$ is the free space characteristic impedance, $\hat{\mathbf{u}}_\varphi$ is the unit vector in the φ direction and I_0 is the amplitude of the current that is assumed to be constant over the small loop. ϑ is the angle with respect to the loop normal, see Figure 6.18.

(a) Calculate the radiated power per unit of solid angle.

(b) Calculate the normalized radiation pattern.

(c) Calculate the radiation resistance. (Hint: Start by calculating the total radiated power).

(d) If the loss resistance is given by

$$R_{loss} = \frac{a}{b} \sqrt{\frac{\pi f \mu_0}{\sigma}}, \tag{6.73}$$

where f is the frequency, a is the loop radius, b is the wire radius and σ is the conductivity of the wire, calculate the antenna radiation efficiency

$$\eta = \frac{R_{rad}}{R_{rad} + R_{loss}}, \tag{6.74}$$

if $a = \frac{\lambda_0}{20}$, $b = \frac{\lambda_0}{2000}$, $\sigma = 5.8 \cdot 10^7 \text{ Sm}^{-1}$ and $f = 1 \text{ GHz}$.

References

1. J.E. Lindsay, Jr., 'A Circular Loop Antenna with Nonuniform Current Distribution', *IRE Transactions on Antennas and Propagation*, pp. 439–441, July 1960.
2. Donald Foster, 'Loop Antennas with Uniform Current', *Proceedings of the IRE*, pp. 603–607, October 1944.
3. David M. Pozar, *Microwave Engineering, second edition*, John Wiley & Sons, New York, 1998.
4. Constantine A. Balanis, *Antenna Theory, Analysis and Design, second edition*, John Wiley & Sons, New York, 1997.
5. John D. Kraus, *Antennas*, McGraw-Hill, New York, 1950.
6. G.N. Watson, *A Treatise on the Theory of Bessel Functions, second edition*, Cambridge at the University Press, Cambridge, UK, 1966.
7. Richard C. Johnson, *Antenna Engineering Handbook, third edition*, McGraw-Hill, New York, 1993.
8. K. Fujimote, A. Henderson, K. Hirasawa and J.R. James, *Small Antennas*, John Wiley & Sons, New York, 1987.
9. R. Faraji-Dana and Y. Chow, 'Edge Condition of the Field and AC Resistance of a Rectangular Strip Conductor', *IEE Proceedings*, Vol. 137, Pt.H., No.2, pp. 133–140, April 1990.
10. Frederick Grover, *Inductance Calculations Working Formulas and Tables*, Dover Publications, 1946.
11. Larry Burgess, 'Matching RFIC Wireless Transmitters to Small Loop Antennas', *High Frequency Electronics*, pp. 20–28, March 2005.

7

Aperture Antennas

We have seen how radiated far magnetic and electric fields can be calculated if the current density on the antenna is known or can be fairly assessed. For small dipole and loop antennas we can assume the current density to be constant. For thin dipole antennas we can – in a first order approximation – assume the current to be sinusoidal. For an increased accuracy, more terms must be added, complicating the analysis. For large loop antennas (circumference larger than a tenth of the free-space wavelength) we can only find a solution easily if we take measures to keep the current density uniform and in-phase. For aperture antennas, like a horn or parabolic reflector antenna, an assessment of the current density over the conducting, three-dimensional structure in general is a far from easy task. Often, however, it is possible to make a fair assessment of the fields in the opening or aperture. In this chapter, we will use these fields to form equivalent sources and calculate the far magnetic and electric fields from these equivalent sources.

When looking at an aperture, like for example the horn antenna shown in Figure 7.1, we immediately see the difficulty in assessing the current density on the metallic parts of the antenna.

An assessment of the fields in the opening or aperture of the horn antenna is fairly easy though. We can imagine a TEM-like pattern in the aperture with straight electric field lines between top and bottom conductor (eventually corrected for the conducting sidewalls) and magnetic field lines perpendicular to the electric field lines. From the fields in the aperture, we will derive equivalent sources. These equivalent sources will consist of an electrical current density and a (fictitious) magnetic current density. We already know how to calculate the radiated fields due to an electric current density. Before we can calculate the fields radiated by an aperture, we need to be able to calculate the radiated fields due to a magnetic current density. When we have accomplished that, we will be able – through the *uniqueness theorem*[1] and the *equivalence principle*.[2] – to calculate the fields radiated by an aperture antenna.

---

[1] The uniqueness theorem, loosely formulated, states that the fields in a volume are determined uniquely by the tangential electric fields over part of the surface and the tangential magnetic fields over the rest of the surface.

[2] The equivalence principle, loosely formulated, relates the fields outside a source volume to equivalent current densities on the surface where the fields are zero inside the source volume.

Antenna Theory and Applications, First Edition. Hubregt J. Visser.
© 2012 John Wiley & Sons, Ltd. Published 2012 by John Wiley & Sons, Ltd.

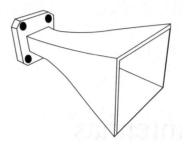

Figure 7.1 Electromagnetic horn antenna.

7.1 Magnetic Sources

For calculating the radiated fields due to (fictitious) magnetic sources, we start with the Maxwell equations (4.11)–(4.14) and remove the electric sources ($\mathbf{J}_e(\mathbf{r}_0) = 0$ and $\rho_e(\mathbf{r}_0) = 0$). Next we introduce the magnetic sources – that is a magnetic current density $\mathbf{J}_m(\mathbf{r}_0)$ and a magnetic charge density $\rho_m(\mathbf{r}_0)$ – without for the moment bothering about the physical meaning, according to

$$\nabla \times \mathbf{E}(\mathbf{r}) = -j\omega\mu_0\mathbf{H}(\mathbf{r}) - \mathbf{J}_m(\mathbf{r}), \tag{7.1}$$

$$\nabla \times \mathbf{H}(\mathbf{r}) = j\omega\varepsilon_0\mathbf{E}(\mathbf{r}), \tag{7.2}$$

$$\nabla \cdot \mathbf{H}(\mathbf{r}) = \frac{\rho_m(\mathbf{r})}{\mu_0}, \tag{7.3}$$

$$\nabla \cdot \mathbf{E}(\mathbf{r}) = 0. \tag{7.4}$$

We have to find $\mathbf{E}$ and $\mathbf{H}$,[3] given the sources $\mathbf{J}_m$ and ρ_m. As in Section 4.2, we will introduce a vector potential to be able, in the end, to calculate the radiated fields from the magnetic current density only.

From equation (7.4) it follows that we may introduce an electric vector potential $\mathbf{A}_m$ through

$$\mathbf{E} = -\frac{1}{\varepsilon_0}\nabla \times \mathbf{A}_m. \tag{7.5}$$

Substituting equation (7.5) in equation (7.2) gives

$$\nabla \times \mathbf{H} = -j\omega\nabla \times \mathbf{A}_m. \tag{7.6}$$

But if equation (7.6) is true, the following equation is also true

$$\nabla \times \mathbf{H} = -j\omega\nabla \times \mathbf{A}_m - \nabla \times \nabla\phi_m, \tag{7.7}$$

since $\nabla \times \nabla\phi_m = 0$, where ϕ_m is an arbitrary scalar. We have added the term $\nabla \times \nabla\phi_m$ to obtain the most general expression for $\mathbf{H}$:

$$\mathbf{H} = -j\omega\mathbf{A}_m - \nabla\phi_m. \tag{7.8}$$

---

[3] From now on, we will stop writing the $\mathbf{r}$-dependence explicitly as long as this does not give rise to confusion.

Substitution of equations (7.5) and (7.8) in equation (7.1) results in

$$-\frac{1}{\varepsilon_0}\nabla \times \nabla \times \mathbf{A}_m = -\omega^2\mu_0\mathbf{A}_m + j\omega\mu_0\nabla\phi_m - \mathbf{J}_m, \tag{7.9}$$

which may be written as

$$\nabla\nabla \cdot \mathbf{A}_m - \nabla^2\mathbf{A}_m = \omega^2\varepsilon_0\mu_0\mathbf{A}_m - j\omega\varepsilon_0\mu_0\nabla\phi_m + \varepsilon_0\mathbf{J}_m. \tag{7.10}$$

Applying the *Lorentz gauge*

$$\phi_m = -\frac{1}{j\omega\varepsilon_0\mu_0}\nabla \cdot \mathbf{A}_m \tag{7.11}$$

to equation (7.10), leads to

$$\nabla^2\mathbf{A}_m + k_0^2\mathbf{A}_m = -\varepsilon_0\mathbf{J}_m, \tag{7.12}$$

a vectorial Helmholtz equation.

Similarly, we find

$$\nabla^2\phi_m + k_0^2\phi_m = -\frac{\rho_m}{\mu_0}, \tag{7.13}$$

a scalar Helmholtz equation.

Then, in a procedure completely analogous to the one described in Section 4.2, we find for the vector potential

$$\mathbf{A}_m(\mathbf{r}) = \frac{\varepsilon_0}{4\pi}\iiint_{V_0} \mathbf{J}_m(\mathbf{r}_0)\frac{e^{-jk_0|\mathbf{r}-\mathbf{r}_0|}}{|\mathbf{r}-\mathbf{r}_0|}dV_0. \tag{7.14}$$

The far electric and magnetic fields are approximated as, following the procedure in Section 4.2,

$$\mathbf{E} \approx \frac{jk_0e^{-jk_0r}}{4\pi r}\hat{\mathbf{u}}_r \times \iiint_{V_0} \mathbf{J}_m(\mathbf{r}_0)e^{jk_0\hat{\mathbf{u}}_r\cdot\mathbf{r}_0}dV_0, \tag{7.15}$$

$$\mathbf{H} \approx -\frac{k_0^2}{j\omega\mu_0}\frac{e^{-jk_0r}}{4\pi r}\hat{\mathbf{u}}_r \times \hat{\mathbf{u}}_r \times \iiint_{V_0} \mathbf{J}_m(\mathbf{r}_0)e^{jk_0\hat{\mathbf{u}}_r\cdot\mathbf{r}_0}dV_0, \tag{7.16}$$

which satisfy

$$\mathbf{E} = Z_0\mathbf{H} \times \hat{\mathbf{u}}_r, \tag{7.17}$$

where $Z_0 = \sqrt{\mu_0/\varepsilon_0}$ is the characteristic impedance of free space.

We now know how to calculate the (far) radiated fields due to electric sources and due to (fictitious) magnetic sources. For mixed sources, we may simply add the contributions due to the linearity of the Maxwell equations.

Before we can proceed with the discussion of equivalent sources – a discussion in which we will give a physical meaning to the magnetic sources – we first need to discuss the so-called uniqueness theorem.

7.2   Uniqueness Theorem

A solution to a problem is said to be unique if this solution is the only possible answer out of a collection of answers. Uniqueness relates fields and the sources that generate these fields. For discussing this, let's assume a source volume V_0, surrounded by a surface S_0, see Figure 7.2. Within the source volume we allow the existence of both electric and magnetic current densities.[4]

Within S_0 we assume the presence of sources $\mathbf{J}_e$ and $\mathbf{J}_m$. The medium within S_0 we assume to be linear, but lossy. This means that

$$\varepsilon = \varepsilon' - j\varepsilon'', \tag{7.18}$$

$$\mu = \mu' - j\mu'' \tag{7.19}$$

All fields within S_0 have to satisfy the Maxwell equations:

$$\nabla \times \mathbf{E}(\mathbf{r}) = -j\omega\mu\mathbf{H}(\mathbf{r}) - \mathbf{J}_m(\mathbf{r}), \tag{7.20}$$

$$\nabla \times \mathbf{H}(\mathbf{r}) = j\omega\varepsilon\mathbf{E}(\mathbf{r}) + \mathbf{J}_e(\mathbf{r}), \tag{7.21}$$

$$\nabla \cdot \mathbf{H}(\mathbf{r}) = \frac{\rho_m(\mathbf{r})}{\mu}, \tag{7.22}$$

$$\nabla \cdot \mathbf{E}(\mathbf{r}) = \frac{\rho_e(\mathbf{r})}{\varepsilon}. \tag{7.23}$$

We now assume two possible solutions within S_0, $(\mathbf{E}_a, \mathbf{H}_a)$ and $(\mathbf{E}_b, \mathbf{H}_b)$. The difference fields are defined by

$$\delta\mathbf{E} = \mathbf{E}_a - \mathbf{E}_b, \tag{7.24}$$

$$\delta\mathbf{H} = \mathbf{H}_a - \mathbf{H}_b. \tag{7.25}$$

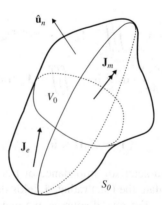

Figure 7.2   Source volume V_0, surrounded by surface S_0, containing electric and magnetic current densities $\mathbf{J}_e$ and $\mathbf{J}_m$. $\hat{\mathbf{u}}_n$ is the outward-directed normal vector.

---

[4] In the figure we have adopted the convention of using single-pointed arrows to indicate electric current densities and double-pointed arrows to indicate magnetic current densities.

Substitution of $\mathbf{E}_a$ and $\mathbf{H}_a$ in equation (7.20), followed by substitution of $\mathbf{E}_b$ and $\mathbf{H}_b$ in equation (7.20) and subtracting the newly formed equations from each other leads to

$$\nabla \times \delta\mathbf{E} = -j\omega\mu\delta\mathbf{H}. \tag{7.26}$$

Substitution of $\mathbf{E}_a$ and $\mathbf{H}_a$ in equation (7.21), followed by substitution of $\mathbf{E}_b$ and $\mathbf{H}_b$ in equation (7.21) and subtracting the newly formed equations from each other leads to

$$\nabla \times \delta\mathbf{H} = j\omega\varepsilon\delta\mathbf{E}. \tag{7.27}$$

The difference fields thus satisfy the source-free Maxwell equations.

We now take the dot product of equation (7.26) with $\delta\mathbf{H}^*$, where $*$ means *complex conjugate*. Then we take the dot product of the complex conjugate of equation (7.27) with $\delta\mathbf{E}$ and subtract this dot product from the first one. The result is

$$\delta\mathbf{E} \cdot (\nabla \times \delta\mathbf{H}^*) - \delta\mathbf{H}^* \cdot (\nabla \times \delta\mathbf{E}) = j\omega\varepsilon\delta\mathbf{E} \cdot \delta\mathbf{E}^* + j\omega\mu\delta\mathbf{H} \cdot \delta\mathbf{H}^*. \tag{7.28}$$

This equation can be simplified, using the vector identity $\nabla \cdot (\mathbf{A} \times \mathbf{B}) = \mathbf{B} \cdot (\nabla \times \mathbf{A}) - \mathbf{A} \cdot (\nabla \times \mathbf{B})$ and using $\mathbf{A} \cdot \mathbf{A}^* = |\mathbf{A}|^2$, into

$$\nabla \cdot (\delta\mathbf{E} \times \delta\mathbf{H}^*) = -j\omega\varepsilon |\delta\mathbf{E}|^2 - j\omega\mu |\delta\mathbf{H}|^2. \tag{7.29}$$

Next, we integrate teh left- and right-hand sides of equation (7.29) over the source volume and apply the divergence (Gauss) theorem to the left-hand side of the resulting equation to obtain:

$$\iint_{S_0} (\delta\mathbf{E} \times \delta\mathbf{H}^*) \cdot \hat{\mathbf{u}}_n dS_0$$
$$= -\iiint_{V_0} \left[\omega\varepsilon'' |\delta\mathbf{E}|^2 + \omega\mu'' |\delta\mathbf{H}|^2 \right] dV_0$$
$$- j \iiint_{V_0} \left[\omega\varepsilon' |\delta\mathbf{E}|^2 + \omega\mu' |\delta\mathbf{H}|^2 \right] dV_0. \tag{7.30}$$

Since in a lossy medium, ε'' and μ'' have to be larger than zero and since ε' and μ' are larger than zero, the condition

$$\iint_{S_0} (\delta\mathbf{E} \times \delta\mathbf{H}^*) \cdot \hat{\mathbf{u}}_n dS_0 = 0, \tag{7.31}$$

can only be satisfied if everywhere inside S_0, $\delta\mathbf{E} = \delta\mathbf{H} = 0$. The same applies for a loss-free medium.

Using the vector identity $\mathbf{A} \cdot \mathbf{B} \times \mathbf{C} = \mathbf{B} \cdot \mathbf{C} \times \mathbf{A} = \mathbf{C} \cdot \mathbf{A} \times \mathbf{B}$, we may write

$$\iint_{S_0} (\delta\mathbf{E} \times \delta\mathbf{H}^*) \cdot \hat{\mathbf{u}}_n dS_0 = \iint_{S_0} (\hat{\mathbf{u}}_n \times \delta\mathbf{E}) \cdot \delta\mathbf{H}^* dS_0 = \iint_{S_0} (\delta\mathbf{H}^* \times \hat{\mathbf{u}}_n) \cdot \delta\mathbf{E} dS_0. \tag{7.32}$$

From equation (7.32) we conclude that condition (7.31) is satisfied if

• the tangential component of the electric field is specified on S_0, $\hat{\mathbf{u}}_n \times \delta\mathbf{E} = 0$ on S_0;

- the tangential component of the magnetic field is specified on S_0, $\hat{\mathbf{u}}_n \times \delta\mathbf{H} = 0$ on S_0;
- the tangential component of the electric field is specified on a part of S_0 and the tangential component of the magnetic field is specified on the remaining part of S_0. $\hat{\mathbf{u}}_n \times \delta\mathbf{E} = 0$ on S_1, $\hat{\mathbf{u}}_n \times \delta\mathbf{H} = 0$ on S_2 and $S_1 \cup S_2 = S_0$.

We will apply these results in the discussion of the *equivalence principle*.

7.3 Equivalence Principle

By using the equivalence principle, the fields outside an (imaginary) closed surface are obtained by defining appropriate electric and magnetic current densities over this surface that satisfy the boundary conditions (continuity of **E**- and **H**-fields over the surface).

In our situation, the closed surface encloses the antenna, that is the sources. We will choose surface current densities in such a way that the fields inside the closed volume become zero and the fields outside the closed surface are identical to the original fields. This is schematically shown in Figure 7.3. The surface current densities then become the equivalent sources for calculating the fields outside the closed surface, that is the radiated fields by the aperture antenna.

For the equivalent problem, shown in Figure 7.3(b) to be valid, the fields inside and outside the closed surface S_0 must satisfy the boundary conditions for the tangential components of the electric and magnetic field. If we call the electric and magnetic fields inside S_0, $\mathbf{E}_1$ and $\mathbf{H}_1$, respectively, then

$$\mathbf{J}_{es} = \hat{\mathbf{u}}_n \times (\mathbf{H} - \mathbf{H}_1)\big|_{\mathbf{H}_1=0} = \hat{\mathbf{u}}_n \times \mathbf{H}, \tag{7.33}$$

$$\mathbf{J}_{ms} = -\hat{\mathbf{u}}_n \times (\mathbf{E} - \mathbf{E}_1)\big|_{\mathbf{E}_1=0} = -\hat{\mathbf{u}}_n \times \mathbf{E}, \tag{7.34}$$

where $\mathbf{J}_{es}$ and $\mathbf{J}_{ms}$ are the electric and magnetic surface current density, respectively, on the closed surface S_0.

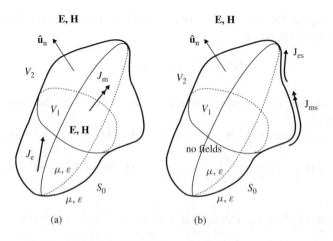

Figure 7.3 Original (a) and equivalent (b) problem.

This form of the equivalence principle is also known as *Love's equivalence principle*.

With the equivalent sources of equations (7.33) and (7.34), the electromagnetic fields may be determined by using superposition and equations (4.43) for the electric sources and equations (7.5)–(7.14) for the magnetic sources. The fields outside S_0 are then found as

$$
\begin{aligned}
\mathbf{E}(\mathbf{r}) &= \frac{1}{j\omega\varepsilon_0\mu_0}\nabla_r \times \nabla_r \times \mathbf{A}_e - \frac{1}{\varepsilon_0}\nabla_r \times \mathbf{A}_m \\
&= \frac{1}{j\omega\varepsilon_0\mu_0}\nabla_r \times \nabla_r \times \frac{\mu_0}{4\pi}\iiint_{V_0} \mathbf{J}_{es}(\mathbf{r}_0)\frac{e^{-jk_0|\mathbf{r}-\mathbf{r}_0|}}{|\mathbf{r}-\mathbf{r}_0|}dV_0 \\
&\quad - \frac{1}{\varepsilon_0}\nabla_r \times \frac{\varepsilon_0}{4\pi}\iiint_{V_0} \mathbf{J}_{ms}(\mathbf{r}_0)\frac{e^{-jk_0|\mathbf{r}-\mathbf{r}_0|}}{|\mathbf{r}-\mathbf{r}_0|}dV_0 \\
&= \nabla_r \times \iint_{S_0}\left[\hat{\mathbf{u}}_n \times \mathbf{E}(\mathbf{r}_0)\right]\varphi dS_0 + \frac{1}{j\omega\varepsilon_0}\nabla_r \times \nabla_r \times \iint_{S_0}\left[\hat{\mathbf{u}}_n \times \mathbf{H}(\mathbf{r}_0)\right]\varphi dS_0,
\end{aligned}
\tag{7.35}
$$

$$
\begin{aligned}
\mathbf{H}(\mathbf{r}) &= \frac{1}{\mu_0}\nabla_r\mathbf{A}_e + \frac{1}{j\omega\varepsilon_0\mu_0}\nabla_r \times \nabla_r \times \mathbf{A}_m \\
&= \frac{1}{\mu_0}\nabla_r \times \nabla_r \times \frac{\mu_0}{4\pi}\iiint_{V_0} \mathbf{J}_{es}(\mathbf{r}_0)\frac{e^{-jk_0|\mathbf{r}-\mathbf{r}_0|}}{|\mathbf{r}-\mathbf{r}_0|}dV_0 \\
&\quad + \frac{1}{j\omega\varepsilon_0\mu_0}\nabla_r \times \frac{\varepsilon_0}{4\pi}\iiint_{V_0} \mathbf{J}_{ms}(\mathbf{r}_0)\frac{e^{-jk_0|\mathbf{r}-\mathbf{r}_0|}}{|\mathbf{r}-\mathbf{r}_0|}dV_0 \\
&= \nabla_r \times \iint_{S_0}\left[\hat{\mathbf{u}}_n \times \mathbf{H}(\mathbf{r}_0)\right]\varphi dS_0 - \frac{1}{j\omega\mu_0}\nabla_r \times \nabla_r \times \iint_{S_0}\left[\hat{\mathbf{u}}_n \times \mathbf{E}(\mathbf{r}_0)\right]\varphi dS_0,
\end{aligned}
\tag{7.36}
$$

where

$$
\varphi = \frac{1}{4\pi}\frac{e^{-jk_0|\mathbf{r}-\mathbf{r}_0|}}{|\mathbf{r}-\mathbf{r}_0|}.
\tag{7.37}
$$

This is known as the *Lorentz–Lamor theorem*. Let's recapitulate:

Lorentz–Lamor Theorem

Assume that within a volume V_0, surrounded by surface S_0 the sources $\mathbf{J}_e$ and $\mathbf{J}_m$ are present. The electromagnetic fields outside V_0 then follow from

$$
\mathbf{E}(\mathbf{r}) = \nabla_r \times \iint_{S_0}\left[\hat{\mathbf{u}}_n \times \mathbf{E}(\mathbf{r}_0)\right]\varphi dS_0 + \frac{1}{j\omega\varepsilon_0}\nabla_r \times \nabla_r \times \iint_{S_0}\left[\hat{\mathbf{u}}_n \times \mathbf{H}(\mathbf{r}_0)\right]\varphi dS_0,
\tag{7.38}
$$

and

$$
\mathbf{H}(\mathbf{r}) = \nabla_r \times \iint_{S_0}\left[\hat{\mathbf{u}}_n \times \mathbf{H}(\mathbf{r}_0)\right]\varphi dS_0 - \frac{1}{j\omega\mu_0}\nabla_r \times \nabla_r \times \iint_{S_0}\left[\hat{\mathbf{u}}_n \times \mathbf{E}(\mathbf{r}_0)\right]\varphi dS_0,
\tag{7.39}
$$

where

$$\varphi = \frac{1}{4\pi} \frac{e^{-jk_0|\mathbf{r}-\mathbf{r}_0|}}{|\mathbf{r} - \mathbf{r}_0|}, \qquad (7.40)$$

and where $\hat{\mathbf{u}}_n \times \mathbf{H} = \mathbf{J}_{es}$ and $\mathbf{E} \times \hat{\mathbf{u}}_n = \mathbf{J}_{ms}$ are the equivalent electric surface current density and equivalent magnetic surface current density, respectively, on the surface S_0. Within the volume V_0, the electric and magnetic fields are zero.

7.4 Radiated Fields

To calculate the far-fields from the equivalent sources, we may apply the same approximations as in Section 4.3 where we approximated the far-fields radiated by an electric current density.

In the integrals in equations (7.38) and (7.39) we encounter products of vector functions and scalar functions. The vector functions do not depend on $\mathbf{r}$, while the operator '$\nabla_r \times$' only operates on $\mathbf{r}$. The scalar function is a function of both $\mathbf{r}_0$, the source point, and $\mathbf{r}$, the observation point. We have seen equations with a similar structure in Section 4.3. There we saw that for observation point $\mathbf{r}$ in the far-field, the vector operations transform into

$$\nabla_r \times \ \rightarrow \ -jk_0\hat{\mathbf{u}}_r \times, \qquad (7.41)$$

$$\nabla_r \times \nabla_r \times \ \rightarrow \ (-jk_0)^2 \, \hat{\mathbf{u}}_r \times \hat{\mathbf{u}}_r \times . \qquad (7.42)$$

In the far-field we also found that, applying different approximations in the phase and amplitude for the vector between source point and observation point,

$$\frac{e^{-jk_0|\mathbf{r}-\mathbf{r}_0|}}{|\mathbf{r} - \mathbf{r}_0|} \approx \frac{e^{-jk_0r}}{r} e^{jk_0\hat{\mathbf{u}}_r \cdot \mathbf{r}_0}. \qquad (7.43)$$

Substitution of equations (7.41)–(7.43) in equations (7.35) and (7.36) then leads to the following results for the far electric and magnetic fields:

$$\mathbf{E}(\mathbf{r}) = \frac{-jk_0 e^{-jk_0r}}{4\pi r}\hat{\mathbf{u}}_r \times \iint_{S_0} \left(\left[\hat{\mathbf{u}}_n \times \mathbf{E}\,(\mathbf{r}_0)\right] - \hat{\mathbf{u}}_r \times \left[\hat{\mathbf{u}}_n \times Z_0\mathbf{H}\,(\mathbf{r}_0)\right] \right) e^{jk_0\hat{\mathbf{u}}_r \cdot \mathbf{r}_0} dS_0,$$

$$(7.44)$$

and

$$Z_0\mathbf{H}(\mathbf{r}) = \frac{-jk_0 e^{-jk_0r}}{4\pi r}\hat{\mathbf{u}}_r \times \iint_{S_0} \left(\left[\hat{\mathbf{u}}_n \times Z_0\mathbf{H}\,(\mathbf{r}_0)\right] + \hat{\mathbf{u}}_r \times \left[\hat{\mathbf{u}}_n \times \mathbf{E}\,(\mathbf{r}_0)\right] \right) e^{jk_0\hat{\mathbf{u}}_r \cdot \mathbf{r}_0} dS_0.$$

$$(7.45)$$

Now that we know how to calculate the far electric and magnetic fields from the tangential fields on a surface, we will apply this first to a rectangular aperture antenna and second to a circular aperture antenna.

7.5   Uniform Distribution in a Rectangular Aperture

Let's assume that we have a horn antenna with a rectangular aperture as shown in Figure 7.4(a) and that the source generating the electromagnetic waves is contained within the waveguide-horn structure as shown in Figure 7.4(b).

We have already mentioned that calculating the far electric and magnetic fields based on an assumed electric current density is very difficult for such an antenna. We have seen though that we may transform the source volume into a source volume without internal fields and having equivalent surface current densities. To use this concept we have to choose the source volume – or more especially the closed surface of the source volume – wisely.

As a start we may assume that, in a first order approximation, the metal walls of the waveguide and horn are perfectly electrically conducting. Therefore, on the surface of the conductors, S_c, see Figure 7.4,

$$\hat{\mathbf{u}}_n \times \mathbf{E}\,(\mathbf{r}_0) = 0 \text{ on } S_c, \tag{7.46}$$

where $\hat{\mathbf{u}}_n$ is the outward-directed normal on the surface S_c.

We may also assume that the currents flowing on the outside of the conductors, that is on S_c, are negligible and therefore

$$\hat{\mathbf{u}}_n \times \mathbf{H}\,(\mathbf{r}_0) = 0 \text{ on } S_c. \tag{7.47}$$

So, if we choose the closed surface S_0 to be flush with the waveguide-horn structure, that is $S_0 = S_c \cup S_a$, where S_a is the surface over the horn opening, the aperture, substitution of equations (7.46) and (7.47) in equations (7.44) and (7.45) shows that for calculating the far-fields $S_0 = S_a$.

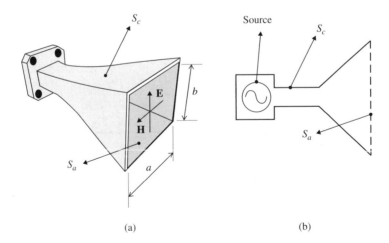

(a)   (b)

Figure 7.4   Horn antenna and chosen surrounding surface. (a) Three-dimensional view. (b) Side view.

So, if we are able to assess the electric and magnetic fields in the horn aperture, we may use equations (7.46) and (7.47) to calculate the far fields.

A fair assessment of the aperture fields would be to take the electric and magnetic fields inside an infinitely long rectangular waveguide with cross-sectional dimensions a and b, see Figure 7.4(a). To keep the calculations simple, however, we choose to use a uniform distribution as would be a fair assessment for a rectangular aperture in an infinite, electrically conducting ground plane. The educational benefit of using this wrong field assessment is considered to outweigh the use of a physically wrong field distribution.[5]

So, the field distribution will be, see also Figure 7.5.

$$\mathbf{E}_0 = E_0\hat{\mathbf{u}}_y \text{ on } S_a, \tag{7.48}$$

$$\mathbf{H}_0 = -\frac{E_0}{Z_0}\hat{\mathbf{u}}_x \text{ on } S_a, \tag{7.49}$$

where S_a is defined by $-\frac{a}{2} \le x \le \frac{a}{2}$ and $-\frac{b}{2} \le y \le \frac{b}{2}$. Furthermore, we have assumed that the characteristic impedance inside the aperture is equal to the characteristic impedance of free space.

Substitution of equation (7.49) in equation (7.33) gives the equivalent electric surface current density as

$$\mathbf{J}_{es} = \hat{\mathbf{u}}_n \times \mathbf{H}_0 = -\frac{E_0}{Z_0}\hat{\mathbf{u}}_z \times \hat{\mathbf{u}}_x = -\frac{E_0}{Z_0}\hat{\mathbf{u}}_y, \tag{7.50}$$

and substitution of equation (7.48) in equation (7.34) gives the equivalent magnetic surface current density as

$$\mathbf{J}_{ms} = -\hat{\mathbf{u}}_n \times \mathbf{E}_0 = -E_0\hat{\mathbf{u}}_z \times \hat{\mathbf{u}}_y = E_0\hat{\mathbf{u}}_x. \tag{7.51}$$

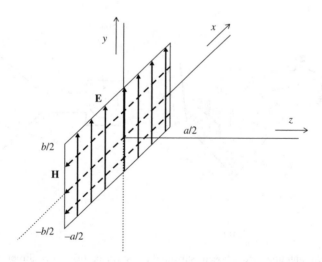

Figure 7.5 Uniform rectangular aperture field distribution.

[5] Once the calculations with the oversimplified assessed aperture field distribution is completed, the reader should be able to recalculate the fields with a proper assessment of the aperture field distribution.

Substitution of equations (7.50) and (7.51) in equation (7.44) then gives, for the far electric field of the horn antenna,

$$
\begin{aligned}
\mathbf{E}(\mathbf{r}) &= \frac{-jk_0 e^{-jk_0 r}}{4\pi r} \hat{\mathbf{u}}_r \times \int_{x_0=-\frac{a}{2}}^{\frac{a}{2}} \int_{y_0=-\frac{b}{2}}^{\frac{b}{2}} \left[-\mathbf{J}_{ms} - Z_0 \hat{\mathbf{u}}_r \times \mathbf{J}_{es} \right] e^{jk_0 \hat{\mathbf{u}}_r \cdot \mathbf{r}_0} dx_0 dy_0 \\
&= \frac{jk_0 e^{-jk_0 r}}{4\pi r} \int_{-\frac{a}{2}}^{\frac{a}{2}} \int_{-\frac{b}{2}}^{\frac{b}{2}} \left[\hat{\mathbf{u}}_r \times \mathbf{J}_{ms} (\mathbf{r}_0) \right] e^{jk_0 \hat{\mathbf{u}}_r \cdot \mathbf{r}_0} dx_0 dy_0 \\
&\quad + \frac{jk_0 e^{-jk_0 r}}{4\pi r} Z_0 \int_{-\frac{a}{2}}^{\frac{a}{2}} \int_{-\frac{b}{2}}^{\frac{b}{2}} \left[\hat{\mathbf{u}}_r \times \hat{\mathbf{u}}_r \times \mathbf{J}_{es} (\mathbf{r}_0) \right] e^{jk_0 \hat{\mathbf{u}}_r \cdot \mathbf{r}_0} dx_0 dy_0 \\
&= \frac{jk_0 e^{-jk_0 r}}{4\pi r} E_0 \int_{-\frac{a}{2}}^{\frac{a}{2}} \int_{-\frac{b}{2}}^{\frac{b}{2}} \left[\hat{\mathbf{u}}_r \times \hat{\mathbf{u}}_x \right] e^{jk_0 \hat{\mathbf{u}}_r \cdot \mathbf{r}_0} dx_0 dy_0 \\
&\quad - \frac{jk_0 e^{-jk_0 r}}{4\pi r} E_0 \int_{-\frac{a}{2}}^{\frac{a}{2}} \int_{-\frac{b}{2}}^{\frac{b}{2}} \left[\hat{\mathbf{u}}_r \times \hat{\mathbf{u}}_r \times \hat{\mathbf{u}}_y \right] e^{jk_0 \hat{\mathbf{u}}_r \cdot \mathbf{r}_0} dx_0 dy_0.
\end{aligned} \tag{7.52}
$$

With equation (5.8) we find that

$$
\hat{\mathbf{u}}_r \times \hat{\mathbf{u}}_x = \hat{\mathbf{u}}_r \times \left[\hat{\mathbf{u}}_\vartheta \cos\vartheta \cos\varphi - \hat{\mathbf{u}}_\varphi \sin\varphi \right] = \hat{\mathbf{u}}_\varphi \cos\vartheta \cos\varphi + \hat{\mathbf{u}}_\vartheta \sin\varphi. \tag{7.53}
$$

With equation (5.9) we find that

$$
\begin{aligned}
\hat{\mathbf{u}}_r \times \hat{\mathbf{u}}_r \times \hat{\mathbf{u}}_y &= \hat{\mathbf{u}}_r \times \hat{\mathbf{u}}_r \times \left[\hat{\mathbf{u}}_r \sin\vartheta \sin\varphi + \hat{\mathbf{u}}_\vartheta \cos\vartheta \sin\varphi + \hat{\mathbf{u}}_\varphi \cos\varphi \right] \\
&= \hat{\mathbf{u}}_r \times \left[\hat{\mathbf{u}}_\varphi \cos\vartheta \sin\varphi - \hat{\mathbf{u}}_\vartheta \cos\varphi \right] = -\hat{\mathbf{u}}_\vartheta \cos\vartheta \sin\varphi - \hat{\mathbf{u}}_\varphi \cos\varphi. \tag{7.54}
\end{aligned}
$$

To calculate the dot product $\hat{\mathbf{u}}_r \cdot \mathbf{r}_0$ we decompose the two vectors in Cartesian coordinates

$$
\begin{aligned}
\hat{\mathbf{u}}_r \cdot \mathbf{r}_0 &= \left[\hat{\mathbf{u}}_x \sin\vartheta \cos\varphi + \hat{\mathbf{u}}_y \sin\vartheta \sin\varphi + \hat{\mathbf{u}}_z \cos\vartheta \right] \cdot \left[\hat{\mathbf{u}}_x x_0 + \hat{\mathbf{u}}_y y_0 + \hat{\mathbf{u}}_z.0 \right] \\
&= x_0 \sin\vartheta \cos\varphi + y_0 \sin\vartheta \sin\varphi = x_0 u + y_0 v, \tag{7.55}
\end{aligned}
$$

where $u = \sin\vartheta \cos\varphi$ and $v = \sin\vartheta \sin\varphi$.

Substitution of equations (7.53)–(7.55) in equation (7.52) gives for the far electric field

$$
\mathbf{E}(\mathbf{r}) = \frac{jk_0 e^{-jk_0 r}}{4\pi r} E_0 \left\{ \hat{\mathbf{u}}_\vartheta \sin\varphi \left[\cos\vartheta + 1 \right] + \hat{\mathbf{u}}_\varphi \cos\varphi \left[\cos\vartheta + 1 \right] \right\}.
$$

$$
\int_{-\frac{a}{2}}^{\frac{a}{2}} \int_{-\frac{b}{2}}^{\frac{b}{2}} e^{jk_0 (x_0 u + y_0 v)} dx_0 dy_0. \tag{7.56}
$$

The integral in equation (7.56) is calculated as

$$
\begin{aligned}
\int_{-\frac{a}{2}}^{\frac{a}{2}} \int_{-\frac{b}{2}}^{\frac{b}{2}} e^{jk_0 (x_0 u + y_0 v)} dx_0 dy_0 &= \int_{-\frac{a}{2}}^{\frac{a}{2}} e^{jk_0 x_0 u} dx_0 \int_{-\frac{b}{2}}^{\frac{b}{2}} e^{jk_0 y_0 v} dy_0 \\
&= \frac{1}{jk_0 u} e^{jk_0 x_0 u} \Big|_{-\frac{a}{2}}^{\frac{a}{2}} \frac{1}{jk_0 v} e^{jk_0 y_0 v} \Big|_{-\frac{b}{2}}^{\frac{b}{2}}
\end{aligned}
$$

$$= a\frac{1}{k_0\frac{a}{2}u}\frac{1}{2j}\left\{e^{jk_0\frac{a}{2}u} - e^{-jk_0\frac{a}{2}u}\right\} b\frac{1}{k_0\frac{b}{2}v}\frac{1}{2j}\left\{e^{jk_0\frac{b}{2}v} - e^{-jk_0\frac{b}{2}v}\right\}$$

$$= a\frac{\sin\left(\frac{k_0au}{2}\right)}{\left(\frac{k_0au}{2}\right)}b\frac{\sin\left(\frac{k_0bv}{2}\right)}{\left(\frac{k_0bv}{2}\right)} = ab\,\text{sinc}\left(\frac{k_0au}{2}\right)\text{sinc}\left(\frac{k_0bv}{2}\right), \qquad (7.57)$$

where $\text{sinc}(x) = \frac{\sin x}{x}$, so that finally we find for the far electric field of a uniformly illuminated rectangular aperture of size $a \times b$:

$$\mathbf{E}(\mathbf{r}) = E_\vartheta \hat{\mathbf{u}}_\vartheta + E_\varphi \hat{\mathbf{u}}_\varphi, \qquad (7.58)$$

where

$$E_\vartheta = \frac{jk_0E_0abe^{-jk_0r}}{4\pi r}\sin\varphi\,[\cos\vartheta + 1]\,\text{sinc}\left(\frac{k_0au}{2}\right)\text{sinc}\left(\frac{k_0bv}{2}\right), \qquad (7.59)$$

$$E_\varphi = \frac{jk_0E_0abe^{-jk_0r}}{4\pi r}\cos\varphi\,[\cos\vartheta + 1]\,\text{sinc}\left(\frac{k_0au}{2}\right)\text{sinc}\left(\frac{k_0bv}{2}\right), \qquad (7.60)$$

and where $u = \sin\vartheta\cos\varphi$, $v = \sin\vartheta\sin\varphi$ and $\text{sinc}(x) = \frac{\sin x}{x}$.

Before continuing with the radiation patterns, we will take a closer look at the sinc function. In Figure 7.6 we show $\text{sinc}(x)$, $\text{sinc}(2x)$ and $\text{sinc}(3x)$ as a function of x.

Figure 7.6 shows that the curve concentrates around $x = 0$ for increasing α. Looking at the expressions for the electric field components, equations (7.59) and (7.60), we may thus conclude that if the aperture gets larger, that is when a and b increase, the radiation

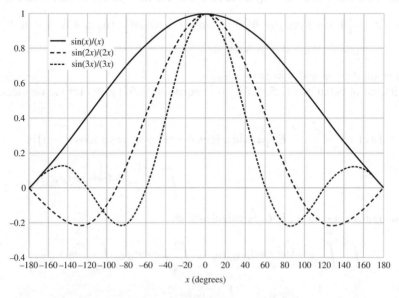

Figure 7.6 $\frac{\sin(\alpha x)}{(\alpha x)}$ as a function of x for $\alpha = 1, 2, 3$.

pattern (that is proportional to the square of the electric field) gets narrower. The beam concentrates in the forward direction, that is in the bore-sight direction.

We may even be more specific. Therefore we will first look at the plane $\varphi = 0$. This is the xz-plane or the horizontal plane in Figures 7.4 and 7.5, the plane parallel to the a-dimension of the aperture.

For $\varphi = 0$, $v = 0$ and $u = \sin \vartheta$. Substitution of these values in equations (7.58)–(7.60) results in

$$\mathbf{E} = \hat{\mathbf{u}}_\varphi \frac{jk_0 E_0 ab e^{-jk_0 r}}{4\pi r} [\cos \vartheta + 1] \operatorname{sinc} \left(\frac{k_0 a \sin \vartheta}{2} \right). \tag{7.61}$$

We see that if we increase a, the beam will become smaller in the $\varphi = 0$ plane, that is the xz-plane.

Next we will look at the plane $\varphi = \frac{\pi}{2}$. This is the yz-plane or the vertical plane in Figures 7.4 and 7.5, the plane parallel to the b-dimension of the aperture.

For $\varphi = \frac{\pi}{2}$, $u = 0$ and $v = \sin \vartheta$. Substitution of these values in equations (7.58)–(7.60) results in

$$\mathbf{E} = \hat{\mathbf{u}}_\vartheta \frac{jk_0 E_0 ab e^{-jk_0 r}}{4\pi r} [\cos \vartheta + 1] \operatorname{sinc} \left(\frac{k_0 b \sin \vartheta}{2} \right). \tag{7.62}$$

We see that if we increase b, the beam will become smaller in the $\varphi = \frac{\pi}{2}$ plane, that is the yz-plane. Note that this is consistent with what we have seen for a thin wire (dipole) antenna.

To calculate the normalized radiation pattern, $F(\vartheta, \varphi) = \frac{P(\vartheta, \varphi)}{P_{max}}$, we use $P(\vartheta, \varphi) = |r^2 \mathbf{S}(\vartheta, \varphi)|$ and $\mathbf{S}(\vartheta, \varphi) = \frac{1}{2Z_0} |\mathbf{E}(\vartheta, \varphi)|^2 \hat{\mathbf{u}}_r$. Since the maximum radiated power is found for $(\vartheta, \varphi) = (0, 0)$, the normalized far-field (power) radiation pattern is found to be

$$F(\vartheta, \varphi) = \frac{|E_\vartheta(\vartheta, \varphi)|^2 + |E_\varphi(\vartheta, \varphi)|^2}{|E_\vartheta(0, 0)|^2 + |E_\varphi(0, 0)|^2}. \tag{7.63}$$

In Figure 7.7 we show radiation pattern cuts in the plane $\varphi = 0$ for three values of the horn a-dimension, $a = \lambda$, $a = 2\lambda$ and $a = 3\lambda$, where λ is the free space wavelength.

The figure shows that the beam gets smaller for a larger aperture. It also shows that the level of the first side lobe seems not to be able to increase beyond a certain level. This is true and the actual level is $-13.26\,\text{dB}$. This level is dictated by the sinc function for large values of $k_0 a / 2$ dominates the $[\cos \vartheta + 1]$ term, see equation (7.62).

Example
Show that the side-lobe level (SLL) cannot exceed a value of $-13.26\,\text{dB}$.

For large arguments of the sinc function or for small angles ϑ, the sinc function will dominate in the calculation of the far electric field. The maximum side-lobe level will then be determined by this sinc function. To find the maximum, we need to calculate the derivative with respect to the argument to zero:

$$\frac{d}{dx} \left(\frac{\sin x}{x} \right) = \cos x \cdot \frac{1}{x} + \sin x \cdot -\frac{1}{x^2} = \frac{x \cos x - \sin x}{x^2} = 0. \tag{7.64}$$

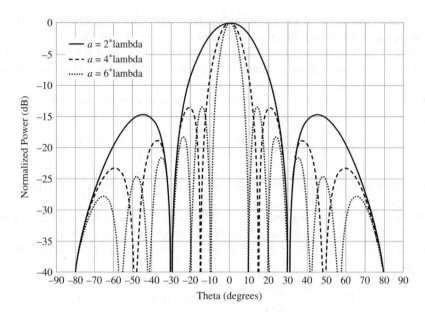

Figure 7.7 Radiation pattern cuts in the plane $\varphi = 0$ for three values of the horn a-dimension.

The first solution, $x = 0$, is obvious and represents the main beam. The second solution may be found by bisectioning. The result is $x \approx 4.5$. The value of the sinc function is found to be $|\sin(4.5)/4.5| = 0.217$ and the corresponding power level is $20 \log(0.217) = -13.26 \, \text{dB}$.

For non-uniform aperture illuminations (as we will actually encounter in a horn aperture) all side lobes will be at a lower level. Aperture weighing is a method for controlling (lowering) the side lobes. The field in the aperture will be distributed in such a way that the side lobes will be at acceptable levels. Whatever measure is intentionally or unintentionally taken, the sinc-dictated SLL of $-13.26 \, \text{dB}$ is a maximum value that cannot be exceeded.

7.6 Uniform Distribution in a Circular Aperture

As an example of a uniform field distribution in a circular aperture antenna, we take the parabolic ('dish') reflector antenna shown in Figure 7.8. A horn antenna illuminates the reflector and the reflected field in the planar, circular aperture S_a will serve as source for the horn-reflector antenna system. The reflector is positioned in the far-field region of the horn antenna.

We suppose that the parabolic reflector is perfect electrically conducting. Furthermore, we assume that the currents on the back of the reflector are negligible. Thus, with reference to Figure 7.8

$$\hat{\mathbf{u}}_n \times \mathbf{E} = 0 \text{ on } S_c, \tag{7.65}$$

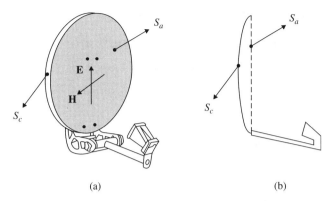

Figure 7.8 Parabolic reflector antenna, illuminated by a small horn antenna. (a) Three-dimensional view. (b) Side view.

and

$$\hat{\mathbf{u}}_n \times \mathbf{H} = 0 \text{ on } S_c. \tag{7.66}$$

Then, choosing the closed, surrounding surface S_0 flush with the parabolic reflector, $S_0 = S_c \cup S_a$, see Figure 7.8, substitution of equations (7.65) and (7.66) in equations (7.44) and (7.45) shows – as in the case with the rectangular aperture antenna – that for calculating the far-fields, $S_0 = S_a$.

The electric and magnetic fields on S_a are the fields originating from the horn antenna, after reflection against the parabolic reflector. In a first order approximation, we may assume these fields to be uniform, see Figure 7.8,

$$\mathbf{E}_0 = E_0 \hat{\mathbf{u}}_x, \tag{7.67}$$

$$\mathbf{H}_0 = \frac{E_0}{Z_0} \hat{\mathbf{u}}_y. \tag{7.68}$$

In Figure 7.8, the x-axis is in the vertical direction, the y-axis is in the horizontal direction and the z-axis is in the forward direction. Substitution of equations (7.67) and (7.68) in equation (7.44) gives

$$\mathbf{E}(\mathbf{r}) = \frac{-jk_0 e^{-jk_0 r}}{4\pi r} \hat{\mathbf{u}}_r \times \iint_{S_0} \left(\left[\hat{\mathbf{u}}_n \times \mathbf{E}_0(\mathbf{r}_0) \right] - \hat{\mathbf{u}}_r \times \left[\hat{\mathbf{u}}_n \times Z_0 \mathbf{H}(\mathbf{r}_0) \right] \right) e^{jk_0 \hat{\mathbf{u}}_r \cdot \mathbf{r}_0} dS_0. \tag{7.69}$$

With $\hat{\mathbf{u}}_n = \hat{\mathbf{u}}_z$ and the aid of equations (7.67) and (5.9), we find that

$$\hat{\mathbf{u}}_r \times \hat{\mathbf{u}}_n \times \mathbf{E}_a = E_0 \hat{\mathbf{u}}_r \times \hat{\mathbf{u}}_z \times \hat{\mathbf{u}}_x = E_0 \hat{\mathbf{u}}_r \times \hat{\mathbf{u}}_y$$
$$= E_0 \hat{\mathbf{u}}_r \times \left[\sin \vartheta \sin \varphi \hat{\mathbf{u}}_r + \cos \vartheta \sin \varphi \hat{\mathbf{u}}_\vartheta + \cos \varphi \hat{\mathbf{u}}_\varphi \right]$$
$$= E_0 \cos \vartheta \sin \varphi \hat{\mathbf{u}}_\varphi - E_0 \cos \varphi \hat{\mathbf{u}}_\vartheta. \tag{7.70}$$

With the aid of equations (7.68) and (5.8) we find that

$$\hat{\mathbf{u}}_r \times \hat{\mathbf{u}}_n \times Z_0\mathbf{H}_a = E_0\hat{\mathbf{u}}_r \times \hat{\mathbf{u}}_z \times \hat{\mathbf{u}}_y = -E_0\hat{\mathbf{u}}_r \times \hat{\mathbf{u}}_x$$

$$= E_0\hat{\mathbf{u}}_r \times \left[\sin\vartheta \cos\varphi\hat{\mathbf{u}}_r + \cos\vartheta \cos\varphi\hat{\mathbf{u}}_\vartheta - \sin\varphi\hat{\mathbf{u}}_\varphi\right]$$

$$= -E_0 \cos\vartheta \cos\varphi\hat{\mathbf{u}}_\varphi - E_0 \sin\varphi\hat{\mathbf{u}}_\vartheta, \tag{7.71}$$

so that

$$\hat{\mathbf{u}}_r \times \hat{\mathbf{u}}_r \times \hat{\mathbf{u}}_n \times Z_0\mathbf{H}_0 = E_0 \cos\vartheta \cos\varphi\hat{\mathbf{u}}_\vartheta - E_0 \sin\varphi\hat{\mathbf{u}}_\varphi, \tag{7.72}$$

and

$$\mathbf{E}_0(\mathbf{r}) = \frac{-jk_0e^{-jk_0r}}{4\pi r}E_0(1 + \cos\vartheta)\left[\sin\varphi\hat{\mathbf{u}}_\varphi - \cos\varphi\hat{\mathbf{u}}_\vartheta\right]\iint_{S_0} e^{jk_0\hat{\mathbf{u}}_r \cdot \mathbf{r}_0}dS_0. \tag{7.73}$$

The exponent in equation (7.73) may be written as, using equations (5.8) and (5.9),

$$\hat{\mathbf{u}}_r \cdot \mathbf{r}_0 = \hat{\mathbf{u}}_r \cdot \left[r_0 \cos\varphi_0\hat{\mathbf{u}}_x + r_0 \sin\varphi_0\hat{\mathbf{u}}_y\right]$$

$$= r_0\hat{\mathbf{u}}_r \cdot \left[\sin\vartheta \cos\varphi \cos\varphi_0\hat{\mathbf{u}}_r + \cos\vartheta \cos\varphi \cos\varphi_0\hat{\mathbf{u}}_\vartheta - \sin\varphi \cos\varphi_0\hat{\mathbf{u}}_\varphi \right.$$

$$\left. + \sin\vartheta \sin\varphi \sin\varphi_0\hat{\mathbf{u}}_r + \cos\vartheta \sin\varphi \sin\varphi_0\hat{\mathbf{u}}_\vartheta + \cos\varphi \sin\varphi_0\hat{\mathbf{u}}_\varphi\right]$$

$$= r_0 \sin\vartheta \cos(\varphi - \varphi_0). \tag{7.74}$$

The surface element dS_0 of the circular aperture may be written as, see also Figure 7.9,

$$dS_0 = rdrd\varphi_0. \tag{7.75}$$

Substitution of equations (7.74) and (7.75) in equation (7.73) gives

$$\mathbf{E}(\mathbf{r}) = \frac{-jk_0e^{-jk_0r}}{4\pi r}E_0(1 + \cos\vartheta)\left[\sin\vartheta\hat{\mathbf{u}}_\varphi - \cos\varphi\hat{\mathbf{u}}_\vartheta\right].$$

$$\int_{r_0=0}^{a}\int_{\varphi_0=0}^{2\pi} e^{jk_0\sin\vartheta \cos(\varphi-\varphi_0)}r_0dr_0d\varphi_0. \tag{7.76}$$

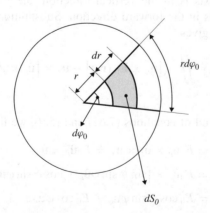

Figure 7.9   The surface element dS_0 of a circular surface is equal to $rdrd\varphi_0$. Not drawn to scale.

By recognizing that [1]

$$\int_0^{2\pi} e^{jk_0 r_0 \sin \vartheta \cos(\varphi - \varphi_0)} d\varphi_0 = 2\pi J_0 (k_0 r_0 \sin \vartheta), \qquad (7.77)$$

where $J_0(x)$ is a Bessel function of the first kind, order zero and argument x, and – after substituting $r_0 = a\rho$ – using [1], we find that

$$\int_{r_0=0}^{a} \int_{\varphi_0=0}^{2\pi} e^{jk_0 \sin \vartheta \cos(\varphi - \varphi_0)} r_0 dr_0 d\varphi_0 = 2\pi \int_{r_0=0}^{a} J_0 (k_0 r_0 \sin \vartheta) r_0 dr_0$$

$$= 2\pi a^2 \int_{\rho=0}^{1} J_0 (k_0 a\rho \sin \vartheta) \rho d\rho = 2\pi a^2 \frac{J_1 (k_0 a \sin \vartheta)}{(k_0 a \sin \vartheta)}, \qquad (7.78)$$

where $J_1(x)$ is a Bessel function of the first kind, order one and argument x.

So, finally, we find for the far electric field

$$\mathbf{E}(\mathbf{r}) = E_\vartheta \hat{\mathbf{u}}_\vartheta + E_\varphi \hat{\mathbf{u}}_\varphi, \qquad (7.79)$$

where

$$E_\vartheta = \frac{ja^2 k_0 E_0 e^{-jk_0 r}}{2r} (1 + \cos \vartheta) \cos \varphi \frac{J_1 (k_0 a \sin \vartheta)}{(k_0 a \sin \vartheta)}, \qquad (7.80)$$

$$E_\varphi = \frac{-ja^2 k_0 E_0 e^{-jk_0 r}}{2r} (1 + \cos \vartheta) \sin \varphi \frac{J_1 (k_0 a \sin \vartheta)}{(k_0 a \sin \vartheta)}. \qquad (7.81)$$

The normalized power radiation pattern is given by

$$F(\vartheta, \varphi) = \frac{|E_\vartheta (\vartheta, \varphi)|^2 + |(\vartheta, \varphi)|^2}{|E_\vartheta (0, 0)|^2 + |E_\varphi (0, 0)|^2}. \qquad (7.82)$$

The numerator of equation (7.82) is found with equations (7.80) and (7.81)

$$|E_\vartheta (\vartheta, \varphi)|^2 + |(\vartheta, \varphi)|^2 = \left[\frac{a^2 k_0 E_0}{2r}\right]^2 (1 + \cos \vartheta)^2 \frac{J_1^2 (k_0 a \sin \vartheta)}{(k_0 a \sin \vartheta)^2}. \qquad (7.83)$$

For calculating the denominator of equation (7.82) we make use of [1]

$$\lim_{x \to 0} \frac{J_1(x)}{x} = \frac{1}{2}, \qquad (7.84)$$

and find

$$|E_\vartheta (0, 0)|^2 + |E_\varphi (0, 0)|^2 = \left[\frac{a^2 k_0 E_0}{2r}\right]^2. \qquad (7.85)$$

The normalized radiation pattern is thus found, upon substitution of equations (7.83) and (7.85) in equation (7.82), to be

$$F(\vartheta, \varphi) = F(\vartheta) = (1 + \cos \vartheta)^2 \left[\frac{J_1 (k_0 a \sin \vartheta)}{(k_0 a \sin \vartheta)}\right]^2. \qquad (7.86)$$

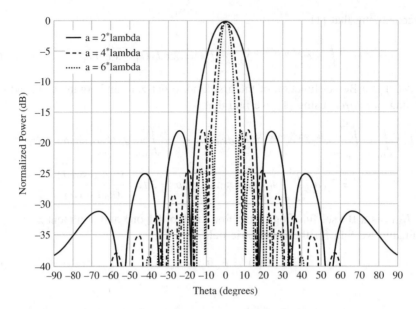

Figure 7.10 Radiation pattern cuts in the plane $\varphi = constant$ for three values of the radius a of a uniformly illuminated circular aperture antenna.

The radiation pattern has become independent of φ. The radiation patterns is rotationally symmetric around the normal on S_a. In Figure 7.10 we show radiation pattern cuts in a plane φ =constant for three values of the a-dimension, $a = 2\lambda$, $a = 4\lambda$ and $a = 6\lambda$, where λ is the free space wavelength.

As in the case with the uniformly illuminated rectangular aperture antenna, we observe that the main beam gets smaller if the aperture gets larger. Furthermore, we see that – for not too broad main beams – the side-lobe level (SLL) remains constant at approximately -17.6 dB. This is due to the dominant behavior of the $J_1(x)/x$ term in equation (7.86).

So, if we want to create a very small antenna beam, we need to apply a very large aperture antenna. For the two antenna types shown, that is the horn antenna and the parabolic reflector antenna, this implies also that the volume of the antenna becomes very large, which is especially impractical if we want to direct the narrow beam in different directions. By 'sampling' the aperture or using an array of small-volume antennas occupying the same physical aperture, we may overcome this impracticality and at the same time provide a means for electronically steering the beam, that is steering the beam without mechanically moving the antenna aperture. This is the subject of the next chapter.

7.7 Microstrip Antennas

An example of a popular aperture antenna is the microstrip patch antenna, although it is not immediately obvious that this type of antenna is indeed an aperture antenna. Figure 7.11 shows a rectangular microstrip patch antenna, which consists of a rectangular metallic patch on top of a grounded dielectric slab.

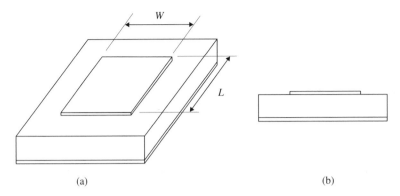

(a)   (b)

Figure 7.11   Rectangular microstrip patch antenna of length L and width W. (a) 3D view. (b) Side view.

The microstrip patch antenna may be thought of as a cavity with electrically conducting bottom and top and magnetically conducting sidewalls. Since it is a cavity, only standing electromagnetic waves, or modes, that 'fit' into the cavity can exist. The first mode will occur when the distance between two opposite boundaries equals half a wavelength. So, while keeping the thickness of the dielectric slab small (in terms of wavelength), making the L-dimension equal to half a wavelength in the dielectric medium will make the structure resonate. For the moment we ignore how to excite this standing wave – we will get to that later. Since the magnetic sidewalls are not perfectly magnetic, the electric field lines will 'fringe' at the patch edges, see Figure 7.12, and this fringing is the origin of the microstrip patch radiation.

The fringe electric fields have vertical and horizontal components. The horizontal electric field components are shown in the top view of the microstrip patch antenna, see Figure 7.12(b). Since the structure is resonant along the L-dimension, we see that the horizontal electric fields along the width of the patch are in phase, while those along the

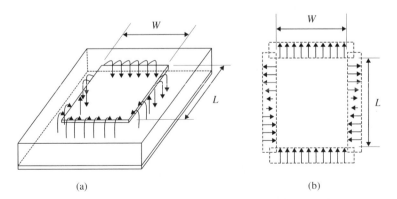

(a)   (b)

Figure 7.12   Fringe electric fields at the edges of a rectangular microstrip patch antenna of length L and width W for the first resonant mode. (a) 3D view. (b) Top view.

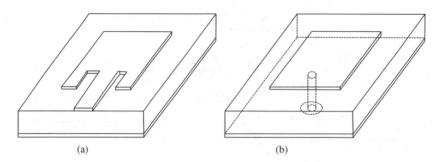

Figure 7.13 Microstrip patch antenna excitation. (a) Inset microstrip transmission line excitation. (b) Probe excitation.

length are in anti-phase. We may consider the microstrip patch antenna as consisting of four small slots [2]. The two slots along the width are, for obvious reasons, called the *radiating slots*, while the slots along the length are known as the *non-radiating slots*. So, indeed, a microstrip patch antenna is an aperture antenna, or, actually, it consists of two aperture antennas.

Excitation of the microstrip patch antenna can be accomplished in a number of ways. Of all the excitation methods available we will only mention the two most used ones. A microstrip transmission line may be used, see Figure 7.13a, where, for a good impedance match, the connection to the patch is moved inward. A probe excitation is also possible, where the outer conductor of a coaxial cable is connected to the ground plane and the inner conductor is guided through the dielectric and connected to the patch, see Figure 7.13(b). The probe position is chosen for a good impedance match.

In the remainder of this section we will employ microstrip transmission line feeding.

7.7.1 *Application of Theory*

With the aid of equations (5.66)–(5.68) we can design a microstrip transmission line of a desired characteristic impedance.

The relation between the resonance frequency, f_0, of the microstrip antenna and the length, L, is given by:

$$f_0 = \frac{1}{2L_{eff}\sqrt{\varepsilon_{reff}}\sqrt{\varepsilon_0\mu_0}}, \tag{7.87}$$

where L_{eff} is the effective length of the microstrip patch antenna and ε_{reff} is the effective relative permittivity of the microstrip patch antenna. The effective length of the patch antenna is given by

$$L_{eff} = L + 2\Delta L, \tag{7.88}$$

where ΔL is a length extension due to the fringing of the field at the edge of the patch antenna. This extension has to be applied twice. This length extension has been empirically determined as [3, 4]

$$\Delta L = 0.412h \frac{\left(\varepsilon_{reff} + 0.3\right)\left(\frac{W}{h} + 0.264\right)}{\left(\varepsilon_{reff} - 0.258\right)\left(\frac{W}{h} + 0.8\right)}, \tag{7.89}$$

where h is the thickness of the dielectric, and the effective relative permittivity may be calculated by [5]

$$\varepsilon_{reff} = \frac{\varepsilon_r + 1}{2} + \frac{\varepsilon_r - 1}{2}\left(1 + \frac{12h}{W}\right)^{\frac{1}{2}}. \tag{7.90}$$

The radiating slot may be modeled, in a first order approximation, as having a uniform field distribution. The far-field is then described by equations (7.58)–(7.60) with a replaced by W and b replaced by h [2].

The antenna now consists of two uniformly excited slots as shown in Figure 7.14

Anticipating the next chapter on array antennas we now have to dwell a bit on the combined radiation/reception of the two slots.

We start by regarding the two-slot system as a receiving antenna. If a plane wave is originating from the far field from the direction of P, see Figure 7.14, the signals received by the two slots will be equal in amplitude but will experience a phase difference. This phase difference is due to the time lag and the associated difference in path lengths between the planar phase front 'hitting' both slots. This is depicted in Figure 7.15 for a plane wave incident from the direction ϑ, $\varphi = \frac{\pi}{2}$.

The phase difference for a wave incident from a general direction ϑ, φ is then given by

$$\psi = k_0(L + \Delta L) \sin\vartheta \sin\varphi, \tag{7.91}$$

where $k_0 = \frac{2\pi}{\lambda_0}$ is the free space wave number, λ_0 being the free space wavelength.

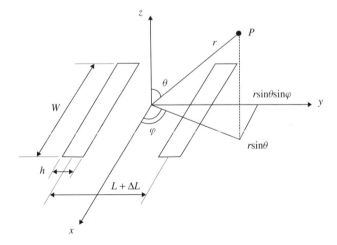

Figure 7.14 Equivalent model of a rectangular microstrip patch antenna.

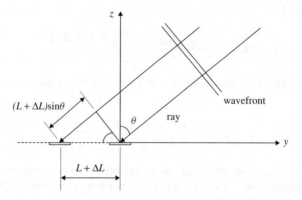

Figure 7.15 Path length difference for a plane wave from the direction ϑ, $\varphi = \frac{\pi}{2}$ incident on a two slot antenna.

In the transmitting situation, the fields from the two slots in the direction ϑ, φ will experience the same phase difference. The normalized far field power radiation pattern is now given by equation (7.63)

$$F(\vartheta, \varphi) = \frac{|E_\vartheta(\vartheta, \varphi)|^2 + |E_\varphi(\vartheta, \varphi)|^2}{|E_\vartheta(0, 0)|^2 + |E_\varphi(0, 0)|^2},\tag{7.92}$$

where

$$E_\vartheta(\vartheta, \varphi) = E_{\vartheta_0}(\vartheta, \varphi) + E_{\vartheta_0}(\vartheta, \varphi)e^{-jk_0(L+\Delta L)\sin\vartheta \sin\varphi}$$

$$= 2E_{\vartheta_0}(\vartheta, \varphi)e^{-jk_0\frac{(L+\Delta L)}{2}\sin\vartheta \sin\varphi}\cos\left[k_0\frac{(L+\Delta L)}{2}\sin\vartheta \sin\varphi\right],\tag{7.93}$$

$$E_\varphi(\vartheta, \varphi) = E_{\varphi_0}(\vartheta, \varphi) + E_{\varphi_0}(\vartheta, \varphi)e^{-jk_0(L+\Delta L)\sin\vartheta \sin\varphi}$$

$$= 2E_{\varphi_0}(\vartheta, \varphi)e^{-jk_0\frac{(L+\Delta L)}{2}\sin\vartheta \sin\varphi}\cos\left[k_0\frac{(L+\Delta L)}{2}\sin\vartheta \sin\varphi\right],\tag{7.94}$$

and

$$E_{\vartheta_0} = j\frac{k_0 E_0 W h}{4\pi r}e^{-jk_0 r}\sin\varphi\,[\cos\vartheta + 1]\frac{\sin\left(\frac{k_0 W \sin\vartheta \cos\varphi}{2}\right)}{\left(\frac{k_0 W \sin\vartheta \cos\varphi}{2}\right)}\frac{\sin\left(\frac{k_0 h \sin\vartheta \sin\varphi}{2}\right)}{\left(\frac{k_0 h \sin\vartheta \sin\varphi}{2}\right)},\tag{7.95}$$

$$E_{\varphi_0} = j\frac{k_0 E_0 W h}{4\pi r}e^{-jk_0 r}\cos\varphi\,[\cos\vartheta + 1]\frac{\sin\left(\frac{k_0 W \sin\vartheta \cos\varphi}{2}\right)}{\left(\frac{k_0 W \sin\vartheta \cos\varphi}{2}\right)}\frac{\sin\left(\frac{k_0 h \sin\vartheta \sin\varphi}{2}\right)}{\left(\frac{k_0 h \sin\vartheta \sin\varphi}{2}\right)}.\tag{7.96}$$

In the above, E_0 is the electric field strength at the rim of the microstrip patch.

From these equations we may conclude that, provided we design a patch antenna to be resonant for the fundamental mode ($L \approx \frac{\lambda_0}{2\sqrt{\varepsilon_r}}$), the radiation pattern will be cosine like and the directivity will be at least 4.8 dB.[6]

7.7.2   Design of a Linearly Polarized Microstrip Antenna

The rectangular microstrip antenna discussed in the previous section is linearly polarized as can be seen by the direction of the electric field in Figure 7.12. We have seen that the length of the patch is roughly equal to half the wavelength in the dielectric substrate. An accurate approximation is obtained by using equations (7.87)–(7.90), but since we will be optimizing an initial design by iteratively using CST-MWS, half the wavelength in the dielectric substrate is a good enough starting value for now.

We will be using a microstrip inset feed, the position of which, together with the width of the microstrip patch antenna, will be used to optimize the impedance matching once the resonance frequency has been tuned by the length of the patch. A good starting value for the width of the microstrip patch antenna is given by [6]

$$W = \frac{\lambda_0}{\sqrt{2\,(\varepsilon_r + 1)}}. \tag{7.97}$$

We now want to design a rectangular, inset feed, microstrip patch antenna, resonant at 2.45 GHz and impedance matched to 50 Ω. The substrate will be FR4, having thickness $h = 1.6$ mm, relative permittivity $\varepsilon_r = 4.28$ and loss tangent $\tan \delta = 0.016$. The top view of the antenna is shown in Figure 7.16.

For the initial values we take the length L equal to half the wavelength in the substrate, leading to $L = 29.6$ mm. For the width W we employ equation (7.97), resulting in $W = 37.7$ mm. With the use of equations (5.66)–(5.68), we find for the microstrip transmission line width $W_f = 3.3$ mm. We fix the gap g, see Figure 7.16, at $g = 1$ mm and for the inset length L_f we choose an arbitrary value that needs to be smaller than $\frac{L}{2}$.[7] We choose $L_f = \frac{L}{4}$.

In analyzing the microstrip patch antenna in CST-MWS, we apply a rim of 15 mm around the patch, attach a waveguide port to the microstrip transmission line, see Figure 7.17, and perform an analysis using the time domain solver.

An adaptive analysis shows a resonance frequency of 2.41 GHz and already a good impedance match, see Figure 7.18. For the impedance match we do not look at the absolute level of the reflection at the resonance frequency but at the -10 dB frequency bandwidth. Any reflection below -10 dB is considered as good, the actual level is of less importance. The broader the -10 dB frequency range is, the better the design will be in compensating for manufacturing tolerances.

---

[6] This assessment is based on the assumption that the slot will, in the worst case, be like a small dipole, thus having a directivity of 1.5, and that decreasing the values of ϑ and φ to zero leads to a multiplicative factor of two, thus yielding a minimum directivity value of 3 or 4.8 dB.

[7] At $\frac{L}{2}$ the electric field underneath the patch equals zero, see Figure 7.12, and the patch thus cannot be excited at that position.

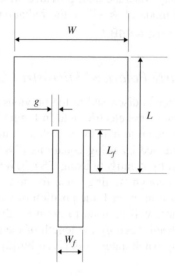

Figure 7.16 Top view of a rectangular, microstrip inset feed excited microstrip patch antenna.

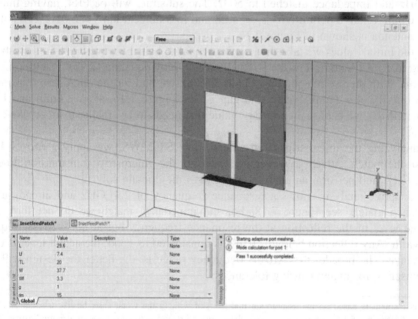

Figure 7.17 Rectangular, microstrip inset feed, microstrip patch antenna as analyzed in CST-MWS.

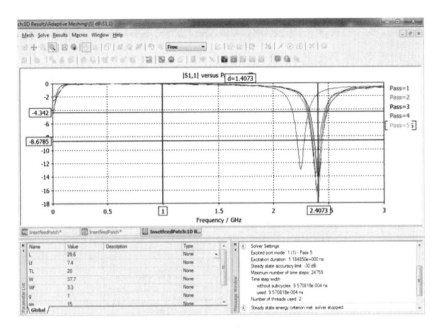

Figure 7.18   Reflection as a function of frequency for the microstrip antenna shown in Figure 7.17.

Before tuning for resonance, we first have a look at the port impedance to check if the characteristic impedance of the microstrip transmission line is 50 Ω. If the port impedance deviates strongly from 50 Ω, the characteristic impedance of the microstrip transmission line will also deviate strongly from this value and the section of transmission line will become part of the antenna. If the characteristic impedance is 50 Ω, the transmission line will not be part of the antenna and will only carry signals from the port to the antenna and vice versa. Figure 7.19 shows the port impedance as a function of the number of analysis passes.

The figure shows that the input impedance is approximately 47.6Ω. This results in a voltage reflection coefficient of

$$\Gamma = \frac{Z_0 - Z_{in}}{Z_0 + Z_{in}} = \frac{50 - 47.6}{50 + 47.6} = 0.025, \qquad (7.98)$$

which is quite acceptable. So we do not need to tune the width of the microstrip transmission line.

Since the resonance frequency is a bit too low, we need to decrease the value of L. Reducing the value to $L = 29.1$ mm results in a matched antenna at the right resonance frequency, see Figure 7.20.

The results are actually so good that we do not see a need to tune the width W or the feed line inset length L_f.

The port impedance value is now approximately 47.8Ω and thus has hardly changed. The radiation is cosine like and shows a directivity of 6.9 dBi, see Figure 7.21.

We may use this linearly polarized microstrip patch antenna as a starting point for designing a circularly polarized microstrip patch antenna.

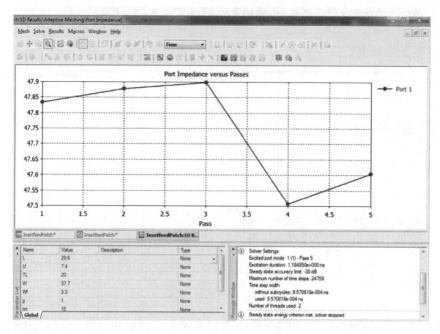

Figure 7.19 Port impedance as a function of the number of analysis passes for the microstrip antenna shown in Figure 7.17.

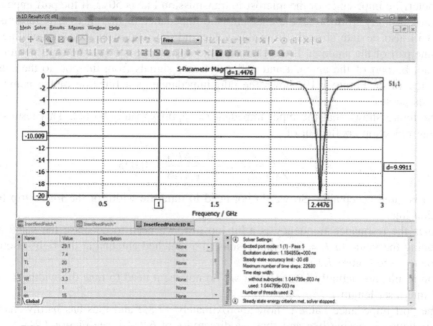

Figure 7.20 Reflection as a function of frequency for the reduced length microstrip antenna.

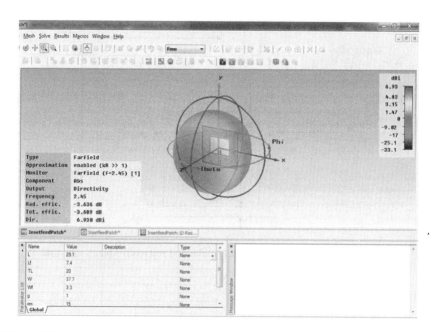

Figure 7.21   Radiation pattern at 2.45 GHz for the reduced length microstrip antenna.

7.7.3   Design of a Circularly Polarized Microstrip Antenna

Circularly polarized waves are employed in, for example, satellite communications and radar. At L-band frequencies, the ionosphere will rotate a linearly polarized wave as it passes through it. By employing circularly polarized waves and antennas, negative effects on linearly polarized waves, such as co-polarized signal attenuation and cross-polarized signal amplification, are avoided. In radar, circularly polarized waves may be employed to 'see through' rain. Upon reflection at a raindrop, a circularly polarized wave will change in direction (right-hand into left-hand and vice versa) and with the correct circular polarization of the receive antenna this reflected signal will not be detected. A linearly polarized wave would shift 180 degrees in phase upon reflection and would still be detected by the linearly polarized receive antenna.

Having thus given a rationale for circularly polarized antennas, we will design a circularly polarized microstrip patch antenna at 2.45 GHz, using the linearly polarized antenna from the previous section as a starting point.

To create circular polarization, we need to excite two orthogonal modes in the patch antenna – at the same frequency – and create a 90 degree phase difference between the two excitations. That means that if we make the width equal to the length of the patch and introduce a second microstrip inset feed at an orthogonal side and introduce the required phase difference, we should be able to create circularly polarized radiation. The required phase difference may be introduced by creating a transmission line path length difference between the two ports that equals a quarter wavelength in the dielectric. The

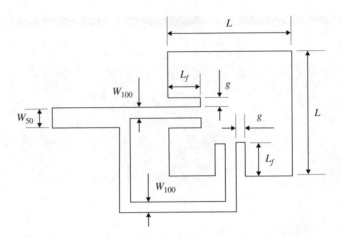

Figure 7.22   Top view of circularly polarized microstrip patch antenna.

phase difference is then given by

$$\psi = k\delta l = \frac{2\pi}{\lambda_g}\frac{\lambda_g}{4} = \frac{\pi}{2}, \tag{7.99}$$

where δl is the transmission line length difference and λ_g is the wavelength in the dielectric.

The top view of the microstrip antenna then looks like shown in Figure 7.22.

In the design, we use a T-splitter to divide the power and introduce a line length difference between the two ports. The characteristic impedance of the input microstrip transmission line is $50\,\Omega$. If we make the characteristic impedances of the microstrip transmission lines going to the patch equal to $100\,\Omega$ and if we design the antenna ports for a $100\,\Omega$ input impedance, the load to the $50\,\Omega$ input transmission line will be two parallel $100\,\Omega$ resistances, which equals $50\,\Omega$.

Using equations (5.66)–(5.68) we find for the width, W_{100}, of a $100\,\Omega$ microstrip transmission line $W_{100} = 0.84\,\text{mm}$, using a 1.6 mm thick FR4 substrate having a relative permittivity $\varepsilon_r = 4.28$ and a loss tangent $\tan\delta = 0.016$. After tuning in CST-MWS we correct this width to $W_{100} = 0.80\,\text{mm}$. The other dimensions of the microstrip patch antenna are taken from the linearly polarized design: $L = 29.1\,\text{mm}$, $L_f = 7.4\,\text{mm}$ and $g = 1\,\text{mm}$. We first analyze the patch antenna with two separate ports to check on the resonance frequency, port impedance and matching, see Figure 7.23.

To start with, we check for the port impedance, see Figure 7.24, which turns out to be approximately $97.5\,\Omega$ for both ports, which is acceptable.

The reflection as a function of frequency at the two ports is shown in Figure 7.25. In the same Figure also the transmission from port to port is shown as a function of frequency. Although full convergence is not yet reached in the results shown, the results are considered good enough for further design.

We do see that the resonance frequency is too low, but that the impedance match (to $100\,\Omega$) is good. We also see from the S_{12} and S_{21} results that the isolation between the two ports is not that high. This must be due to the fact that the two excitation positions

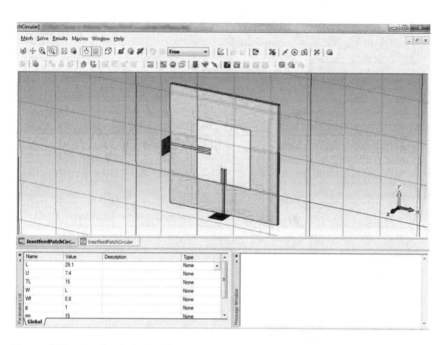

Figure 7.23 Dual polarized microstrip patch antenna with $100\,\Omega$ transmission lines.

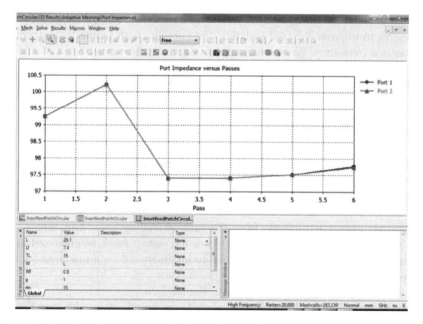

Figure 7.24 Port impedances as a function of analysis passes for the structure shown in Figure 7.23.

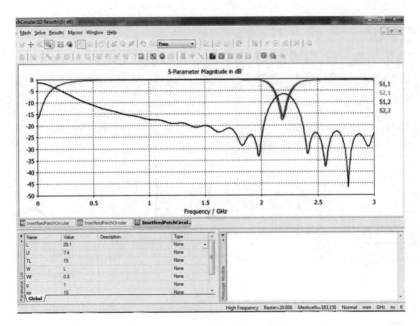

Figure 7.25 Reflection and transmission as a function of frequency at and between the ports of the structure shown in Figure 7.23.

at the patch antenna are positioned very close to each other. For the moment, we do not take measures to improve the port isolation and continue with tuning the antenna for the right resonance frequency. For this we need to decrease the value of L.

With a few iterations of CST-MWS, we find that for resonance $L = 26.2$ mm is a good value. For a good impedance match, the length of the inset slot, that is the feeding location, has been decreased to $L_f = 6$ mm, see Figure 7.26. The scattering parameters as a function of frequency are shown in Figure 7.27.

With this two-port antenna design as a basis, we can now set up the structure for circular polarization as shown in Figure 7.22. Whether the shown structure is for left-hand or right-hand circular polarization is of little importance at the moment. Rotating the feeding structure over 90 degrees counterclockwise will change the polarization direction, so the correct polarization direction can be chosen after finishing the initial design.

In the design process we may expect that we have to tune the microstrip patch dimensions again due to coupling effects from the various feeding lines. This tuning could be avoided by applying a wide separation between microstrip patch antenna and microstrip transmission lines. But since that would lead to an antenna design too large for practical use, we accept the expected tuning.

The layout of the final design is shown in Figure 7.28.

The microstrip patch has been slightly altered, the length and width, see Figure 7.22 have changed into $L = W = 26.2$ mm and the inset feed gap length has become $L_f = 6.0$ mm. The gap between microstrip line edge and the edge of the feeding gap has been increased to 2 mm. The length of the 0.8 mm wide 100 Ω microstrip line from 50 Ω line to the edge of the microstrip patch is 6.4 mm, the (horizontal) length of the bottom

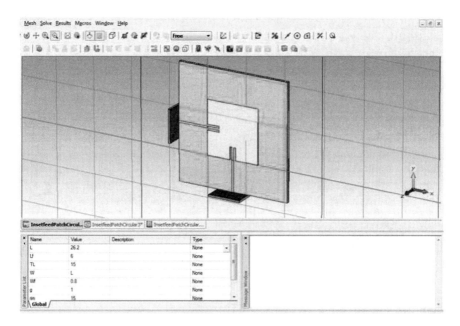

Figure 7.26   Redesigned dual polarized microstrip patch antenna with $100\,\Omega$ transmission lines.

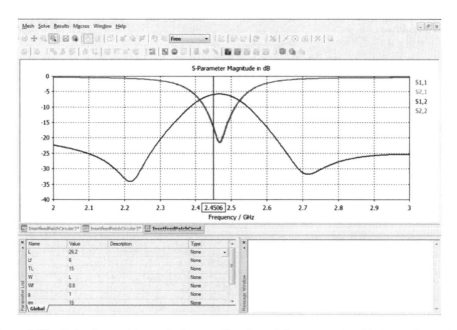

Figure 7.27   Reflection and transmission as a function of frequency at and between the ports of the structure shown in Figure 7.26.

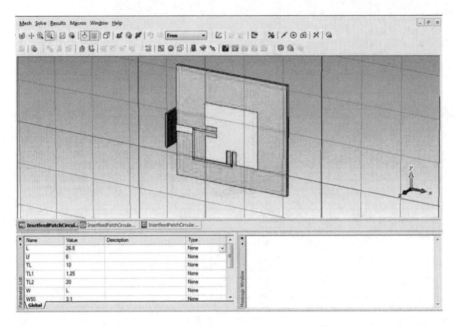

Figure 7.28 Circularly polarized microstrip patch antenna.

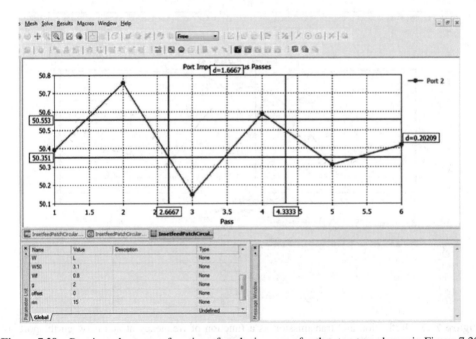

Figure 7.29 Port impedance as a function of analysis passes for the structure shown in Figure 7.28

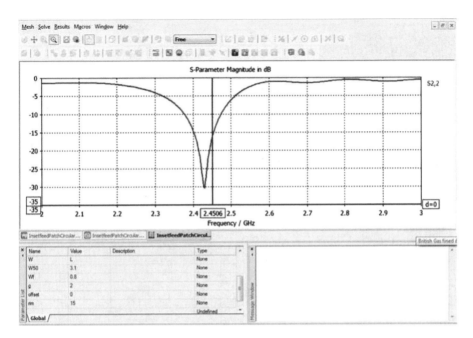

Figure 7.30   Reflection coefficient as a function of frequency for the antenna shown in Figure 7.28

100 Ω microstrip line is 19.80 mm. The distance between the edge of this line and the bottom edge of the microstrip patch antenna is 0.45 mm. The length of the vertical 100 Ω microstrip line is 14.65 mm.

In the tuning process, we have also changed the width of the '50 Ω' microstrip line to 3.1 mm to obtain a closer match to 50 Ω, see Figure 7.29.

Finally, in the next three figures, we show the results of the design. To start with, in Figure 7.30 we show the reflection coefficient as a function of frequency.

Although the resonance is not exactly at 2.45 GHz, the −10 dB frequency bandwidth is wide enough to cover the frequency of interest. In Figure 7.31 the right-hand circularly polarized radiation pattern is shown, demonstrating a 5.68 dBi RHCP directivity at 2.45 GHz.

The cross polarized pattern, that is the orthogonally polarized pattern or LHCP pattern is shown in Figure 7.32, demonstrating a −0.75 dBi directivity.

7.8   Problems

7.1 The far electric field of a (uniformly illuminated) rectangular aperture antenna, having dimensions a and b, see Figure 7.33, is given by

$$\mathbf{E}(\vartheta, \varphi) = E_\vartheta \hat{\mathbf{u}}_\vartheta + E_\varphi \hat{\mathbf{u}}_\varphi, \tag{7.100}$$

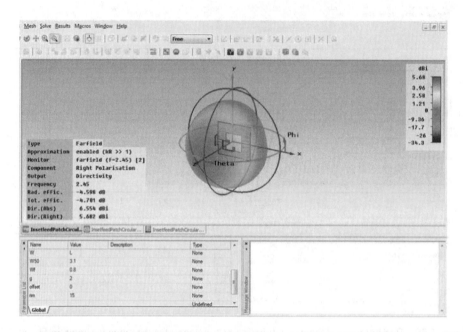

Figure 7.31   RHCP radiation pattern for the antenna shown in Figure 7.28.

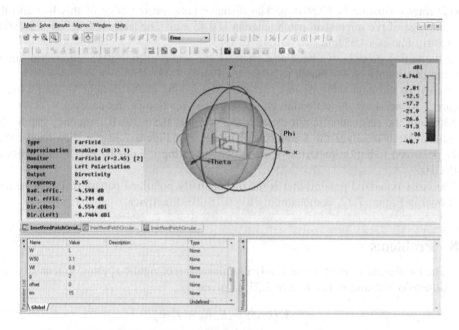

Figure 7.32   LHCP radiation pattern for the antenna shown in Figure 7.28.

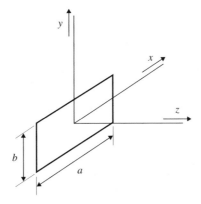

Figure 7.33 Rectangular aperture antenna.

where $\hat{\mathbf{u}}_\vartheta$ and $\hat{\mathbf{u}}_\varphi$ are unit vectors in the ϑ- and φ-directions, respectively and where

$$E_\vartheta = \frac{jk_0 E_0 a b e^{-jk_0 r}}{4\pi r} \sin\varphi \left[\cos\vartheta + 1\right] \cdot$$

$$\frac{\sin\left(\frac{k_0 a}{2} \sin\vartheta \cos\varphi\right)}{\left(\frac{k_0 a}{2} \sin\vartheta \cos\varphi\right)} \frac{\sin\left(\frac{k_0 b}{2} \sin\vartheta \sin\varphi\right)}{\left(\frac{k_0 b}{2} \sin\vartheta \sin\varphi\right)}, \qquad (7.101)$$

$$E_\varphi = \frac{jk_0 E_0 a b e^{-jk_0 r}}{4\pi r} \cos\varphi \left[\cos\vartheta + 1\right] \cdot$$

$$\frac{\sin\left(\frac{k_0 a}{2} \sin\vartheta \cos\varphi\right)}{\left(\frac{k_0 a}{2} \sin\vartheta \cos\varphi\right)} \frac{\sin\left(\frac{k_0 b}{2} \sin\vartheta \sin\varphi\right)}{\left(\frac{k_0 b}{2} \sin\vartheta \sin\varphi\right)}. \qquad (7.102)$$

Here, $k_0 = \frac{2\pi}{\lambda_0}$ is the free space wave number.

(a) Derive an expression for the normalized radiation pattern.

(b) For $\varphi = \frac{\pi}{2}$ (yz-plane), plot, between $\vartheta = -\frac{\pi}{2}$ and $\vartheta = \frac{\pi}{2}$, the normalized radiation pattern on a logarithmic scale for $b = 2\lambda_0$.

(c) Quantify from this graph the half power beamwidth.

(d) For $\varphi = \frac{\pi}{2}$ (yz-plane), plot, between $\vartheta = -\frac{\pi}{2}$ and $\vartheta = \frac{\pi}{2}$, the normalized radiation pattern on a logarithmic scale for $b = \frac{\lambda_0}{2}$.

(e) Quantify from this graph the half power beamwidth.

7.2 Calculate the theoretical maximum value of the second side lobe relative to the main beam for a uniformly illuminated rectangular aperture antenna.

7.3 For the rectangular microstrip patch antenna shown in Figure 7.21, the dimensions are $W = 37.7\,mm$, $L = 29.1$ mm and $h = 1.6$ mm. Use these dimensions as input for equations (7.92)–(7.96) to calculate the normalized radiation pattern in

(a) the plane $\varphi = 0$ (xz-plane).

(b) the plane $\varphi = \frac{\pi}{2}$ (yz-plane).

References

1. Milton Abramowitz and Irene A. Stegun, *Handbook of Mathematical Functions*, Dover Publications, New York, 1972.
2. P. Hammer, D. van Bouchate, D. Verschraeven and A. van de Capelle, 'A Model for Calculating the Radiation Field of Microstrip Antennas', *IEEE Transactions on Antennas and Propagation*, Vol. AP-27, No. 2, pp. 267–270, March 1979.
3. E.O. Hammerstad, 'Equations for Microstrip Circuit Design', *Proceedings European Microwave Conference*, pp. 268–272, September 1975.
4. Constantine A. Balanis, *Antenna Theory, Analysis and Design, second edition*, John Wiley & Sons, New York, 1997.
5. David M. Pozar, *Microwave Engineering, second edition*, John Wiley & Sons, New York, 1998.
6. C.W. Garvin, Robert E. Munson, L.T. Oswald and Klaus G. Schroeder, 'Missile Base Mounted Microstrip Antennas', *IEEE Transactions on Antennas and Propagation*, Vol. AP-25, No. 5, pp. 604–610, September 1977.

8

Array Antennas

We have seen in the discussion of the radiated fields of wire and aperture antennas that the radiated fields consist of contributions from an infinite number of elementary sources. On a macroscopic level, we may group elementary sources into antenna elements and then use these antenna elements to create a larger aperture antenna. The benefits of such an operation are twofold. First an aperture antenna is created occupying a relatively small volume allowing control over the field distribution. Second, by applying a phase-taper over the elements that make up the aperture, the radiated beam may be pointed into a desired direction without physically moving the antenna. Although array antennas come in many forms, including linear, planar, curved and three-dimensional ones, we will only discuss array antenna basics and therefore will stick to the linear array antenna.[1]

8.1 A Linear Array of Non-Isotropic Point-Source Radiators

Let's assume a system of identical radiators, placed at equidistant positions along a straight line, see Figure 8.1. This system is a so-called *linear array antenna*. Neither the identicalness of the radiators, nor their equidistant positions are prerequisites for a linear array antenna. They are introduced though to avoid obscuring the explanation of the linear array antenna basics. Besides, in practical situations, linear array antennas are often realized with identical radiators that are equidistantly positioned.

Also for clarity reasons, we assume that the individual radiators that make up the array antenna do not occupy any volume, but we allow them to have a non-isotropic radiation pattern. In other words, we assume the radiators to be (physically not realizable) *non-isotropic point sources*. Thereby we allow the radiators to have directivity, that is have a non-trivial radiation pattern, but for the moment we do not have to bother with restrictions in positioning the elements due to their own physical dimensions since we simply assume these physical dimensions to be non-existent.

Furthermore, for explaining the array antenna basics, we consider the array antenna to be a receiving antenna. This is not a restriction, since by virtue of the *reciprocity theorem*,

---

[1] The material discussed in this chapter is to a large extend taken from [1].

Antenna Theory and Applications, First Edition. Hubregt J. Visser.
© 2012 John Wiley & Sons, Ltd. Published 2012 by John Wiley & Sons, Ltd.

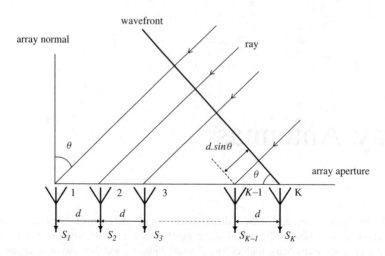

Figure 8.1 A linear array of K radiators, equidistantly positioned along a straight line, where a plane wave is incident at an angle ϑ with respect to the array normal.

we know that the characteristics of an antenna when used as a transmitting antenna are identical to those when used as a receiving antenna.

Accepting these restrictions, we assume, see Figure 8.1, that the wavefront of a plane wave is incident upon the linear array antenna at an angle ϑ to the array aperture. The wavefront is perpendicular to the direction of the plane wave. This direction is indicated by *rays* in the figure.

A wavefront is defined by the characteristic that all points on the wavefront have equal amplitude and phase values.

Our linear array antenna consists of K radiators or elements. At a given moment, the planar wavefront has reached element K, see Figure 8.1. To reach element $K - 1$, the planar wavefront must travel a distance $d \sin(\vartheta)$, as may be seen from the same figure. To reach element $K - 2$, the wavefront must travel a distance $2d \sin(\vartheta)$, and so on. If we normalize the phases of the received signals such that the phase at element K is zero, the phase *differences* with respect to element K of the signals received by the other elements represent the received phases of these elements. These phases, Φ_i, are obtained by multiplying the path lengths by the free space wave number, k_0.

$$\Phi_i = k_0(K - i)d \sin(\vartheta) \text{ for } i = 1, 2, \ldots, K, \tag{8.1}$$

where

$$k_0 = \frac{2\pi}{\lambda_0}. \tag{8.2}$$

Here, λ_0 is the wavelength in free space.

By the term *received signal*, we mean the current flowing through the clamps of the element or the amplitude of the guided wave traveling through the waveguide connected to the element.[2] The exact nature of this signal is of no concern for the explanation of the array antenna basics.

The *complex* signals received by the elements of the array antenna, $S_i(\vartheta)$, may be written as

$$S_i(\vartheta) = S_e(\vartheta) a_i e^{jk_0(K-i)d\sin(\vartheta)} \text{ for } i = 1, 2, \dots, K, \qquad (8.3)$$

where $S_e(\vartheta)$ represents the complex radiation pattern of one isolated individual radiator and a_i is the amplitude received by the ith element. For the moment we will assume that all amplitudes received by the elements are equal and normalized to one, that is

$$a_i = 1 \text{ for } i = 1, 2, \dots, K. \qquad (8.4)$$

We call this a *uniform aperture distribution*.

If we combine all received signals without introducing additional phase differences between the elements, we may simply add the received signals described by equation (8.3) for all elements i. The total received signal, $S(\vartheta)$, is then found to be

$$S(\vartheta) = \sum_{i=1}^{K} S_i(\vartheta) = S_e(\vartheta) \sum_{i=1}^{K} e^{jk_0(K-i)d\sin(\vartheta)}. \qquad (8.5)$$

Combining the received signal without introducing additional phase differences may be accomplished by using a feeding or summing network (in the appropriate waveguide technology) that ensures equal path lengths to all elements of the array. Such a feeding network is schematically shown in Figure 8.2.[3]

In writing down equation (8.5), we have implicitly assumed that the radiation pattern of an individual radiator remains the same upon placing this radiator in an array environment. Apart from the interaction introduced by the inter-element distances and

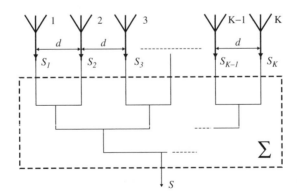

Figure 8.2 Linear array antenna with equal path length summing network.

---

[2] The waveguide may be a hollow rectangular waveguide, a two-wire or a coaxial transmission line, for example.
[3] This type of feeding arrangement is known as a *corporate feeding network*.

the phase differences that these distances cause, no further interaction between the radiators is assumed. In other words we neglect *mutual coupling* between the radiators. Although situations may arise where this assumption is valid, in general it is not true. However, for the explanation of array antenna basics we neglect mutual coupling effects for the moment.

Having accepted this, we return to equation (8.5) and see that the received signal may be separated in a component due to a single radiator and in a component due to the *array configuration* only,

$$S(\vartheta) = S_e(\vartheta)S_a(\vartheta),\tag{8.6}$$

where $S_e(\vartheta)$ is known as the *element factor* and

$$S_a(\vartheta) = \sum_{i=1}^{K} e^{jk_0(K-i)d\sin(\vartheta)}\tag{8.7}$$

is known as the *array factor*.

The element factor is the radiation pattern of a single radiator, and the array factor is the radiation pattern of an array of K *isotropic* radiators. The radiation pattern of the linear array antenna, $S(\vartheta)$, is obtained by multiplying the element factor, $S_e(\vartheta)$, with the array factor, $S_a(\vartheta)$. This operation is known as *pattern multiplication*.

Example
Consider a linear array antenna consisting of eight elements. The element *voltage* radiation pattern is given by

$$S_e(\vartheta) = \cos(\vartheta).\tag{8.8}$$

Given this hypothetical voltage radiation pattern,[4] calculate and show the element factor power pattern, the array factor power pattern and the power radiation pattern of the total array as a function of the angle ϑ relative to the array normal (broadside) for the following element distances d:

1. $d = \frac{\lambda_0}{4}$;
2. $d = \frac{\lambda_0}{2}$;
3. $d = \lambda_0$;
4. $d = \frac{5\lambda_0}{4}$.

Using equations (8.6), (8.7) and (8.8) produces the radiation power patterns shown in Figures 8.3, 8.4, 8.5 and 8.6 for $d = \frac{\lambda_0}{4}$, $d = \frac{\lambda_0}{2}$, $d = \lambda_0$ and $d = \frac{5\lambda_0}{4}$, respectively. The element power pattern is calculated as $20\log(|S_e(\vartheta)|)$, the normalized array factor power pattern is calculated as $20\log(|S_a(\vartheta)|/8)$ and the normalized power pattern of the total array is calculated as $20\log(|S_e(\vartheta)||S_a(\vartheta)|/8)$.

---

[4] Although the element radiation pattern is a hypothetical one, it bears a strong resemblance with the radiation pattern of a horizontal half-wave dipole antenna as may be seen by comparing the (power) radiation pattern with the ones shown for a vertical half-wave dipole antenna. The pattern also resembles that of a slot in an infinite ground plane. Assuming these latter elements, we only need to consider the radiation pattern in the upper hemisphere, $-90° \leq \vartheta \leq 90°$. The power radiation pattern is the square of the voltage radiation pattern.

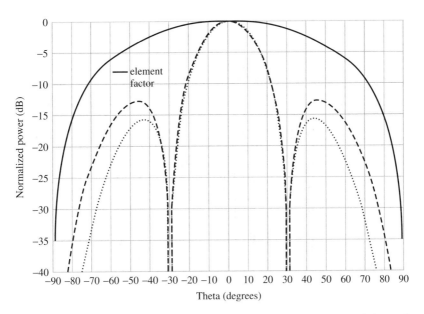

Figure 8.3 Power radiation patterns of the element factor, the array factor and the total array of a linear eight-element broadside array with element distance $d = \frac{\lambda_0}{4}$.

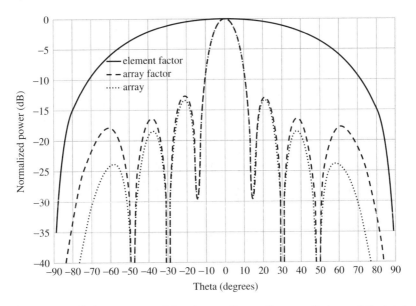

Figure 8.4 Power radiation patterns of the element factor, the array factor and the total array of a linear eight-element broadside array with element distance $d = \frac{\lambda_0}{2}$.

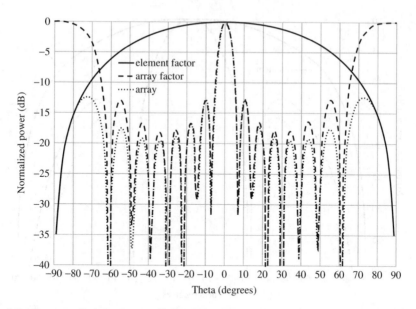

Figure 8.5 Power radiation patterns of the element factor, the array factor and the total array of a linear eight-element broadside array with element distance $d = \lambda_0$.

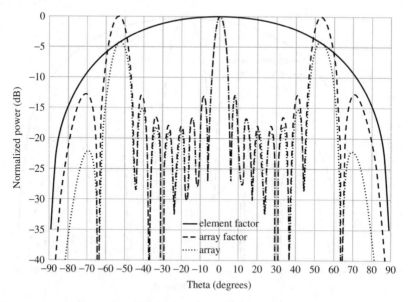

Figure 8.6 Power radiation patterns of the element factor, the array factor and the total array of a linear eight-element broadside array with element distance $d = \frac{5\lambda_0}{4}$.

We clearly see that the main beam of the linear array antenna gets smaller when the elements occupy a larger area, that is when the element distance increases. Furthermore, we see that after passing a critical element distance, a further increase of the element distance leads to the introduction of additional main beams.

We also see that the total linear array antenna behavior is dominated by the array factor. The directive properties of the elements merely act as an angular filter that lowers the radiated power of the array antenna for angles getting closer to *endfire*, that is, directions parallel to the linear array antenna. Due to the dominant character of the array factor we will discuss this array factor in more detail.

8.2   Array Factor

The array factor is given by equation (8.7), which can be rewritten, in more compact form, as

$$S_a(\vartheta) = \sum_{i=1}^{K} e^{jk_0(K-i)d\sin(\vartheta)} = \sum_{i=1}^{K} e^{j(K-i)T}, \tag{8.9}$$

where

$$T = k_0 d \sin(\vartheta). \tag{8.10}$$

In equation (8.9) we may recognize a finite geometric series.

If we multiply both sides of this equation with e^{jT}, we get

$$S_a(\vartheta)e^{jT} = e^{jKT} + e^{j(K-1)T} + \cdots + e^{j2T} + e^{jT}. \tag{8.11}$$

Next, we subtract equation (8.9) from equation (8.11) and thus obtain

$$S_a(\vartheta)\left(e^{jT} - 1\right) = \left(e^{jKT} - 1\right), \tag{8.12}$$

which may be written, after splitting and rearranging the exponential terms, as

$$S_a(\vartheta) = \frac{e^{j\frac{KT}{2}}\left(e^{j\frac{KT}{2}} - e^{-j\frac{KT}{2}}\right)}{e^{j\frac{T}{2}}\left(e^{j\frac{T}{2}} - e^{-j\frac{T}{2}}\right)} = e^{j\frac{K-1}{2}T}\frac{\sin\left(\frac{K}{2}T\right)}{\sin\left(\frac{1}{2}T\right)}, \tag{8.13}$$

so that, finally, using $k_0 = 2\pi/\lambda_0$

$$|S_a(\vartheta)| = \left|\frac{\sin\left(\frac{K}{2}k_0 d\sin(\vartheta)\right)}{\sin\left(\frac{1}{2}k_0 d\sin(\vartheta)\right)}\right| = \left|\frac{\sin\left(\pi\frac{Kd}{\lambda_0}\sin(\vartheta)\right)}{\sin\left(\pi\frac{d}{\lambda_0}\sin(\vartheta)\right)}\right|. \tag{8.14}$$

8.3   Side Lobes and Grating Lobes

With the newly found expression for the (voltage) radiation pattern of a linear array antenna consisting of isotropic radiators, we may now analyze some characteristics of *side lobes* and *grating lobes*.

8.3.1 Side-Lobe Level

Equation (8.14) shows that the maximum of the (voltage) array factor occurs for $\vartheta = 0$ and is equal to K. For angles ϑ close to broadside, that is around the main beam, we may approximate the array factor of equation (8.14) using an approximation for the sine function for small arguments, $\sin(x) \approx x$. This leads to the following expression for the absolute value of the array factor near broadside

$$|S_a(\vartheta)| \approx K \left| \frac{\sin(Kx)}{Kx} \right|, \qquad (8.15)$$

where

$$x = \pi \frac{d}{\lambda_0} \sin(\vartheta). \qquad (8.16)$$

We have specifically chosen this representation, with a factor K in both the numerator and denominator.

We have seen, in discussing the rectangular aperture antenna, that the second maximum of $\left| \frac{\sin(Kx)}{Kx} \right|$ appears for the argument being equal to approximately $Kx \approx 4.5$ and that the amplitude of this maximum is equal to 0.21723.

Therefore the level of the radiation power pattern's first side lobe will be approximately $20 \log 0.21723 = -13.26$ dB, relative to the pattern maximum.

Comparing this value with the first side-lobe levels of the array factor as shown in Figures 8.3–8.6 indicates the validity of our approximation for the array factor in the vicinity of broadside.

8.3.2 Grating Lobes

Equation (8.14) not only shows that the maximum of the (voltage) array factor (K) occurs for $\vartheta = 0$, but the equation also shows that the array factor is a periodic function of ϑ and that multiple maxima occur whenever

$$\pi \frac{d}{\lambda_0} \sin(\vartheta) = m\pi \quad \text{for } m = 1, 2, 3, \ldots, \qquad (8.17)$$

that is for the argument of the sine function being an integer multiple of π. Note that due to the *absolute* value of the sine function appearing in the expression for the radiation pattern, the periodicity has become π instead of 2π.

From the above equation, a restriction for the element distance follows that ensures that there is only one maximum ($m = 1$) within the range $-\vartheta_{max} \leq \vartheta \leq \vartheta_{max}$

$$\frac{d}{\lambda_0} \leq \frac{1}{|\sin(\vartheta_{max})|}. \qquad (8.18)$$

If we do not want to have secondary maxima or lobes within the whole angular range, $-\frac{\pi}{2} \leq \vartheta \leq \frac{\pi}{2}$, the restriction becomes more severe:

$$\frac{d}{\lambda_0} \leq \frac{1}{\left|\sin\left(\frac{\pi}{2}\right)\right|} = 1. \qquad (8.19)$$

If this condition is not met, more than one maximum will occur. These extra maxima are known as *grating lobes*, a term originating from optics.

In our example, grating lobes are occurring for the situation where $d = \lambda_0$, see Figure 8.5, and for the situation where $d = \frac{5\lambda_0}{4}$, see Figure 8.6. Figure 8.5 reveals that keeping too strictly to the grating lobe condition as stated in equation (8.19) does not prevent grating lobe effects being present at all in the radiation pattern. Choosing an inter-element distance too close to one wavelength, still results in unwanted effects in the radiation pattern due to the *slopes* of the first grating lobes. Thus, keeping the grating lobe *maximum* out of visible space ($-\frac{\pi}{2} \leq \vartheta \leq \frac{\pi}{2}$) alone is not enough to ensure that we don't see any grating lobe effects. Therefore, the inter-element distance should be chosen with great care and should be *smaller* than one wavelength.

We need to be aware of the fact that a *grating lobe* differs from an ordinary *side lobe*. Side lobes are the result of constructive and destructive interference from different radiating parts of the antenna. The level of a side lobe is always below that of the main beam. A grating lobe is due to the *periodicity* in the radiation pattern and is formed in directions where a maximum in-phase addition of radiated fields occur. A grating lobe should be compared with the main beam instead of to an ordinary side lobe. The level of a grating lobe is equal to that of the main beam, since a grating lobe is a repeated main beam. As can be seen from the array factor in Figures 8.5 and 8.6, each grating lobe is accompanied by its own set of side lobes that are – just like the grating lobe itself – copies of the radiation pattern around broadside.

Due to the earlier mentioned 'angular filtering characteristics' of the element factor, the grating lobe phenomenon seems to be less severe in Figure 8.5 than in Figure 8.6. In the first figure, the element factor greatly reduces the amplitude of the grating lobe. As we can see in both figures, the side-lobe levels are also affected (lowered) by the element radiation pattern.

By weighing the amplitudes of the array antenna elements we may further control the side-lobe levels. By introducing additional phase differences between the elements, we can change the directive properties of the array antenna.

8.4 Linear Phase Taper

As in the previous chapters we start by considering a linear array antenna consisting of K elements, equally spaced by a distance d. The direction of a wave is described by the angle ϑ between rays and the array normal. The difference with the previous situation is that now, in the (corporate) feed network, we add a microwave two-port between every antenna element and its branch of the feed network, see Figure 8.7.

The transfer function, $H_i(\vartheta)$, of the microwave two-port i, $i = 1, 2, \ldots, K$, is given by

$$H_i(\vartheta) = \frac{S_i(\vartheta)}{S_i'(\vartheta)} = a_i e^{j\psi_i}. \tag{8.20}$$

How to realize such a two-port is beyond the scope of this book. Details may be found in [1]. It suffices to say now that the two-port will allow us to change the amplitude of every received signal $S_i'(\vartheta)$ and – more important for the moment – it will allow us

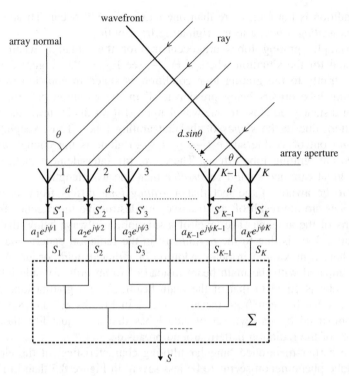

Figure 8.7 A linear *phased* array of K radiators, equidistantly positioned along a straight line, where a plane wave is incident at an angle ϑ with respect to the array normal.

to change the *phase* of the received signal. The two-ports open up the opportunity of operating a *phased array antenna*.

Since $S_i(\vartheta) = S_i'(\vartheta) H_i(\vartheta)$ and we already know from the theory of the linear broadside array antenna that $S_i'(\vartheta) = S_e(\vartheta) e^{jk_0(K-i)d\sin(\vartheta)}$, where $k_0 = \frac{2\pi}{\lambda}$ is the free space wave number and $S_e(\vartheta)$ is the element radiation pattern, we may now write for the array radiation pattern, see also Figure 8.7,

$$S(\vartheta) = \sum_{i=1}^{K} S_i(\vartheta) = S_e(\vartheta) \sum_{i=1}^{K} a_i e^{j[k_0(K-i)d\sin(\vartheta)+\psi_i]}. \tag{8.21}$$

In this equation we have implicitly assumed that mutual coupling effects between the array antenna elements are negligible (or identical for all elements), allowing for a common element radiation pattern that is taken out of the summation.

All the coefficients a_i form an amplitude taper. In order not to obscure the *phased* array antenna discussion, we assume a uniform, normalized amplitude distribution:

$$a_i = 1 \text{ for } i = 1, 2, \ldots, K. \tag{8.22}$$

Now, we only have to consider the array factor, $S_a(\vartheta)$, that is given by

$$S_a(\vartheta) = \sum_{i=1}^{K} e^{j[k_0(K-i)d\sin(\vartheta)+\psi_i]}.$$ (8.23)

If, next, we choose a linear phase taper that is equal to

$$\psi_i = -k_0(K-i)d\sin(\vartheta_0) \text{ for } i = 1, 2, \ldots, K,$$ (8.24)

where $-90° \leq \vartheta_0 \leq 90°$, the array factor may be written as

$$S_a(\vartheta) = \sum_{i=1}^{K} e^{jk_0(K-i)d[\sin(\vartheta)-\sin(\vartheta_0)]}.$$ (8.25)

For $\vartheta_0 = 0$, the phase taper is zero or non-existent and we encounter the linear broadside array antenna situation. The maximum of the array factor was encountered in that situation for $\vartheta = 0$, or – more precisely – for $\sin(\vartheta) = 0$.

For the linear phased array antenna situation we now find the array factor maximum for

$$\sin(\vartheta) - \sin(\vartheta_0) = 0,$$ (8.26)

or, provided that $-90° \leq \vartheta, \vartheta_0 \leq 90°$, for $\vartheta = \vartheta_0$.

So, by choosing a desired beam-pointing direction ϑ_0 and subsequently phasing the linear array antenna elements according to $\psi_i = -k_0(K-i)d\sin(\vartheta_0)$, the array factor will have its maximum at the desired angle $\vartheta = \vartheta_0$.

Example
Consider a linear array antenna consisting of eight elements. The element voltage radiation pattern is given by

$$S_e(\vartheta) = \cos(\vartheta).$$ (8.27)

Given this hypothetical radiation pattern for an aperture element in an infinite ground plane, calculate and show the element power radiation pattern, the array factor power pattern and the power radiation pattern of the total array as a function of the angle ϑ relative to the array normal (broadside) for the following element distances d:

1. $d = \frac{\lambda_0}{4}$;
2. $d = \frac{\lambda_0}{2}$;
3. $d = \lambda_0$;
4. $d = \frac{5\lambda_0}{4}$,

for a phasing aimed at a beam pointing to $\vartheta_0 = 30°$.

Using equations (8.21), (8.25) and (8.27) results in the radiation power patterns shown in Figures 8.8–8.10, and 8.11 for, respectively, $d = \frac{\lambda_0}{4}$, $d = \frac{\lambda_0}{2}$, $d = \lambda_0$ and $d = \frac{5\lambda_0}{4}$. The

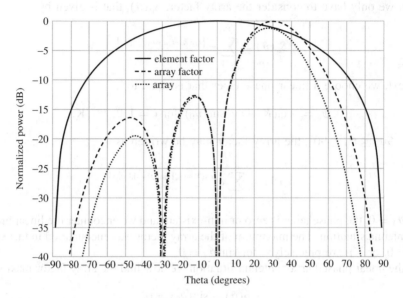

Figure 8.8 Power radiation patterns of the element factor, the array factor and the total array of a linear eight-element phased array antenna with element distance $d = \frac{\lambda_0}{4}$, phased for beam pointing at $\vartheta_0 = 30°$.

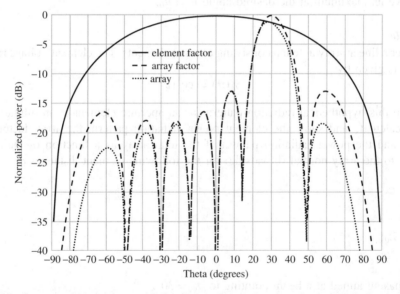

Figure 8.9 Power radiation patterns of the element factor, the array factor and the total array of a linear eight-element phased array antenna with element distance $d = \frac{\lambda_0}{2}$, phased for beam pointing at $\vartheta_0 = 30°$.

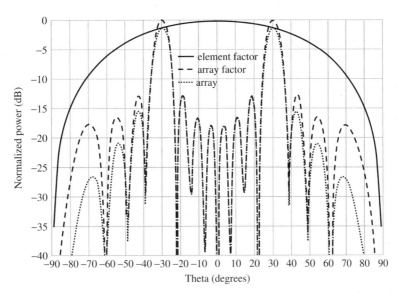

Figure 8.10   Power radiation patterns of the element factor, the array factor and the total array of a linear eight-element phased array antenna with element distance $d = \lambda_0$, phased for beam pointing at $\vartheta_0 = 30°$.

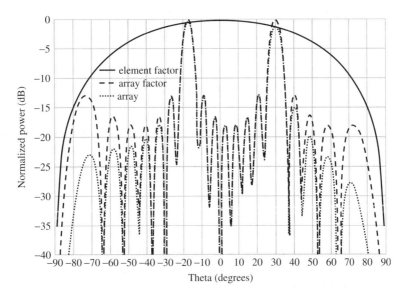

Figure 8.11   Power radiation patterns of the element factor, the array factor and the total array of a linear eight-element phased array antenna with element distance $d = \frac{5\lambda_0}{4}$, phased for beam pointing at $\vartheta_0 = 30°$.

element power pattern is calculated as $20 \log (|S_e(\vartheta)|)$, the normalized array factor power pattern is calculated as $20 \log (|S_a(\vartheta)|/8)$ and the normalized power pattern of the total array is calculated as $20 \log (|S_e(\vartheta)| |S_a(\vartheta)|/8)$.

As expected, the main beam in all situations points in the desired $\vartheta_0 = 30°$ direction. For an element distance of a quarter of a wavelength (Figure 8.8), the aperture size of the array is rather small, resulting in a broad main beam and – although the array factor points at exactly $30°$, the filtering effect of the element pattern means that the maximum of the linear array main beam is at an angle ϑ slightly less than $30°$. We see that upon increasing the element distance of the array, the beam gets narrower and, consequently, the pointing of the beam gets more accurate. The narrowing of the beam is entirely due to the fact that with the increase of the element distance the aperture size increases and, since beamwidth is inversely proportional to aperture size, the beam gets narrower.

We also see that for an element distance of one wavelength and a beam directed towards $\vartheta_0 = 30°$, grating lobes are already well within the visible range ($-90° \leq \vartheta \leq 90°$). Apparently, another condition applies for avoiding grating lobes in a linear *phased* array antenna as compared to a linear broadside array antenna.

8.5   Grating Lobes

When we take a closer look at the array factor again, equation (8.25), we see that whenever

$$k_0 d [\sin \vartheta - \sin \vartheta_0] = n2\pi, \tag{8.28}$$

where n is an integer, the array factor repeats itself. The main beam is identified by $n = 0$, and all other integer values of n identify grating lobes. The first grating lobe ($n = 1$) satisfies

$$\frac{d}{\lambda} = \frac{1}{(\sin \vartheta - \sin \vartheta_0)}, \tag{8.29}$$

where use has been made of $k_0 = \frac{2\pi}{\lambda}$.

When a grating lobe (maximum) just appears at $\vartheta = \pm 90°$, the corresponding $\sin \vartheta$ value is ± 1. The above equation for the first grating lobe then tells us that grating lobes may be just or just not avoided if the element distance relative to the wavelength satisfies

$$\frac{d}{\lambda} \leq \frac{1}{1 + |\sin \vartheta_{0max}|}, \tag{8.30}$$

where ϑ_{0max} is the maximum scan angle. The equality sign applies to the 'just' or 'just not' situation. For the equality sign, the maximum of the grating lobe is present at $\vartheta = \pm 90°$. As with the non-scanning array antenna, for a scanning array antenna the grating lobe condition should be applied more restrictively if also the slope of the first grating lobe should be suppressed.

For $\vartheta_{0max} = 0$ (no scanning at all), we find $\frac{d}{\lambda} \leq 1$. This is the situation we have already encountered for the broadside linear array antenna. If, on the other hand, we do not want a grating lobe maximum to be present in visible space for all possible scan angles, we should take $\vartheta_{0_{max}} = \pm 90°$, which will lead to

$$\frac{d}{\lambda} \leq \frac{1}{2}. \tag{8.31}$$

This explains the grating lobes we have seen for a scanning antenna in the example for element distances exceeding half a wavelength.

When looking at the radiation patterns as a function of the angle ϑ relative to broadside, we see – upon enlarging the element distance – grating lobes appearing at $\vartheta = \pm 90°$. Upon further increasing the element distance, or scanning the beam further to endfire, these grating lobes move from endfire direction closer to broadside direction. The range $-90° \leq \vartheta \leq 90°$, corresponding to $|\sin(\vartheta)| \leq 1$, is called the *visible region*. Grating lobes enter the visible region coming from the *invisible region*.

8.6 Special Topics

In this section we will briefly discuss some topics that are beyond the array antenna basics but that are important enough to be mentioned. We notice that (phased) array antenna technology has evolved and is being transferred from the military and space world into the consumer world and its products. Therefore, we will discuss some aspects related to small and low-cost array antennas, such as diversity and sequential rotation techniques. Before we do, we will first discuss *mutual coupling*, a phenomenon we have left out so far in order not to obscure the discussion.

8.6.1 Mutual Coupling

So far we have been neglecting mutual coupling effects. As apparent from equations (8.5) and (8.6) for example, we have assumed that within an array every element maintains the radiation pattern it would have when completely isolated. In reality, this may be a fair approximation when the antenna elements are not positioned too close together. Theoretically and in practice, for closely spaced elements, this is not true however. When two antennas are brought close together they will interact. This interaction or *mutual coupling*, depends on the separation of the two elements, the orientation or polarization and on the radiation characteristics.

When one or both antennas are transmitting, energy radiated by one of the antennas will induce a current on the other antenna. This induced current will change the input impedance of that antenna and, through reradiation, will change the radiation pattern of that antenna. So in a finite array antenna,[5] even if this array antenna consists of identical elements, each element will have a different input impedance and will demonstrate different radiation characteristics. Mutual coupling effects depend heavily on the type of radiator(s) being employed in the array antenna. Therefore, we cannot derive generalized analytical expressions for the coupling effects. What we can do, however, is present a qualitative description of the coupling effects [1].

---

[5] A very large array antenna of identical elements may be considered as being infinite. In an infinite array, every element experiences an identical environment and thus for every element the mutual coupling effects are identical. For such an array, the principle of pattern multiplication can be used again, using the array element pattern [1]. An array antenna is considered very large if the number of internal element outweighs the number of edge elements.

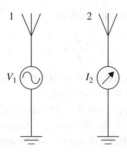

Figure 8.12 Two antennas close to each other. Antenna 1 is transmitting, antenna 2 is receiving.

Consider two antennas that do not have to be identical. We attach a voltage generator V_1 to antenna 1 and measure the current I_2 induced in antenna 2 at the terminals of this antenna, see Figure 8.12.

The ratio of voltage V_1 to current I_2, which has the dimension of impedance, we call the *transfer impedance* or *mutual impedance*

$$\frac{V_1}{I_2} = Z_{12}. \tag{8.32}$$

Similarly, for the situation where antenna 1 is receiving while antenna 2 is transmitting, see Figure 8.13, we find

$$\frac{V_2}{I_1} = Z_{21}. \tag{8.33}$$

By virtue of antenna reciprocity, see Chapter 4,

$$Z_{21} = Z_{12}. \tag{8.34}$$

We now consider the situation where we only have one antenna present – let's say antenna 1, excited by voltage generator V_1. We measure the current I_1 at its terminals and define the input impedance as

$$Z_{11} = \frac{V_1}{I_1}. \tag{8.35}$$

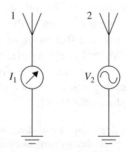

Figure 8.13 Two antennas close to each other. Antenna 1 is receiving, antenna 2 is transmitting.

We now bring antenna 2 into the neighborhood of antenna 1. Through radiation[6] a current I_2 is induced in antenna 2 by antenna 1. This current I_2 will cause (additional) radiation from antenna 2 that will influence the current in antenna 1. This phenomenon is known as *mutual coupling*.

The total voltage at antenna 1 may now be written as

$$V_1 = Z_{11}I_1 + Z_{12}I_2, \tag{8.36}$$

where Z_{11} and Z_{12} are defined as before:

$$Z_{11} = \left. \frac{V_1}{I_1} \right|_{I_2=0}, \tag{8.37}$$

$$Z_{12} = \left. \frac{V_1}{I_2} \right|_{I_1=0}. \tag{8.38}$$

Similarly

$$V_2 = Z_{21}I_1 + Z_{22}I_2, \tag{8.39}$$

where

$$Z_{21} = \left. \frac{V_2}{I_1} \right|_{I_2=0}, \tag{8.40}$$

$$Z_{22} = \left. \frac{V_2}{I_2} \right|_{I_1=0}. \tag{8.41}$$

We may generalize these equations for a two-element array antenna to a K-element array antenna, resulting in:

$$\begin{aligned} V_1 &= Z_{11}I_1 + Z_{12}I_2 + \cdots + Z_{1K}I_K \\ V_2 &= Z_{21}I_2 + Z_{22}I_2 + \cdots + Z_{2K}I_K \\ &\;\;\vdots \\ V_K &= Z_{K1}I_1 + Z_{K2}I_2 + \cdots + Z_{KK}I_K \end{aligned}, \tag{8.42}$$

where

$$Z_{mn} = \left. \frac{V_m}{I_n} \right|_{I_i=0, i \neq n}. \tag{8.43}$$

The input impedance Z_m of the mth element in the array, accounting for all the mutual coupling, is then given by

$$Z_m = \frac{V_m}{I_m} = Z_{m1}\frac{I_1}{I_m} + Z_{m2}\frac{I_2}{I_m} + \cdots + Z_{mn} + \cdots + Z_{mK}\frac{I_K}{I_m}. \tag{8.44}$$

This impedance is also known as the *active impedance*.

---

[6] And in real situations, it will be through scattering from the feeding network of an array or scattering from nearby objects.

Since alternating current is the origin of electromagnetic radiation, the mutual coupling will affect not only the input impedances of the elements in the array, but also their radiation patterns. The mutual coupling effects will in general change with element position, frequency and angle of radiation and will depend on the radiating element under consideration. Thus, in general, the calculation of mutual coupling will be complicated and will require the use of numerical techniques.

As an example we will show how the mutual coupling and radiation will change if we decrease the distance between three identical microstrip patch elements in a linear array antenna. The analysis will be performed by using CST-MWS.

Example

Consider a linear array of three identical microstrip patch antennas, resonant at 2.45 GHz, see Figure 8.14.

The patches have a length (y-direction) of 29.1 mm and a width (x-direction) of 37.7 mm. The microstrip feeding line has a width of 3.3 mm, the inset feed gap length is 7.4 mm and the width is 1 mm. The material of the patch is copper, having a thickness of 70 μm. The grounded substrate is 1.6 mm thick FR4, having a relative permittivity of 4.28 and a loss tangent of 0.016. The spacing between the patches (right side patch to left side subsequent patch) is 60 mm.

We will investigate the mutual coupling by looking at the scattering parameters rather than the impedance parameters. Scattering parameters, like impedance parameters, describe the characteristics of an electrical network. But instead of using open and short circuits to do so, it uses matched terminations. These terminations are easier to realize at microwave frequencies. The scattering parameters describe transmission and reflection

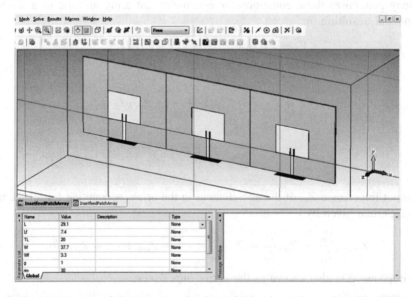

Figure 8.14 Linear array of three rectangular, inset feed, microstrip patch antennas as analyzed in CST-MWS.

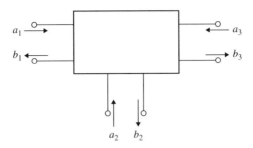

Figure 8.15 In- and outgoing waves in a three-port network.

then in terms of wave amplitude and phase and can be measured directly. S-parameters can be easily converted into Z-parameters (or Y-parameters) and the other way around, see Appendix E.

We will describe the system as the three-port network shown in Figure 8.15. The complex amplitudes of the ingoing waves are denoted a_i, and those of the outgoing waves are denoted b_i, where $i = 1, 2, 3$. The outgoing and ingoing waves are related to each other through the scattering parameters S_{ij}, $i, j = 1, 2, 3$.

$$\begin{pmatrix} b_1 \\ b_2 \\ b_3 \end{pmatrix} = \begin{pmatrix} S_{11} & S_{12} & S_{13} \\ S_{21} & S_{22} & S_{23} \\ S_{31} & S_{32} & S_{33} \end{pmatrix} \cdot \begin{pmatrix} a_1 \\ a_2 \\ a_3 \end{pmatrix}. \tag{8.45}$$

The scattering parameters are defined by

$$S_{11} = \left.\frac{b_1}{a_1}\right|_{a_2=a_3=0} \quad S_{12} = \left.\frac{b_1}{a_2}\right|_{a_1=a_3=0} \quad S_{13} = \left.\frac{b_1}{a_3}\right|_{a_1=a_2=0}$$

$$S_{21} = \left.\frac{b_2}{a_1}\right|_{a_2=a_3=0} \quad S_{22} = \left.\frac{b_2}{a_2}\right|_{a_1=a_3=0} \quad S_{23} = \left.\frac{b_2}{a_3}\right|_{a_1=a_2=0} \tag{8.46}$$

$$S_{31} = \left.\frac{b_3}{a_1}\right|_{a_2=a_3=0} \quad S_{32} = \left.\frac{b_3}{a_2}\right|_{a_1=a_3=0} \quad S_{33} = \left.\frac{b_3}{a_3}\right|_{a_1=a_2=0}$$

S_{11} is therefore the reflection coefficient at port 1 when port 2 and 3 are not excited, S_{22} is the reflection coefficient at port 2 when port 1 and 3 are not excited, S_{12} is the transfer from port 2 to port 1 and so on.

The Z-matrix may be obtained from the S-matrix through

$$[Z] = [G]^{-1} \cdot ([I] - [S])^{-1} \cdot ([S] + [I]) \cdot [Z_{ch}] \cdot [G], \tag{8.47}$$

where $[Z_{ch}]$ is a diagonal matrix with characteristic impedances at the ports of the network

$$[Z_{ch}] = \begin{bmatrix} Z_{ch1} & 0 & 0 \\ 0 & Z_{ch2} & 0 \\ 0 & 0 & Z_{ch3} \end{bmatrix}, \tag{8.48}$$

where $[G]$ is the diagonal matrix

$$[G] = \begin{bmatrix} \frac{1}{\sqrt{Re|Z_{ch1}|}} & 0 & 0 \\ 0 & \frac{1}{\sqrt{Re|Z_{ch2}|}} & 0 \\ 0 & 0 & \frac{1}{\sqrt{Re|Z_{ch3}|}} \end{bmatrix}, \qquad (8.49)$$

and where $[I]$ is the identity matrix

$$[I] = \begin{bmatrix} 1 & 0 & 0 \\ 0 & 1 & 0 \\ 0 & 0 & 1 \end{bmatrix}. \qquad (8.50)$$

For the array antenna shown in Figure 8.14, the scattering parameters as a function of frequency are shown in Figure 8.16.

The figure shows a minimum reflection (S_{11}, S_{22} and S_{33}) – that is, the best matched impedance – at the three antenna ports at the operational frequency. The coupling between adjacent antennas (S_{12}, S_{21}, S_{23} and S_{32}) is very low (< -30dB) and the coupling between the side elements (S_{13} and S_{31}) is even lower (< -40dB). As a result of this low coupling, the radiation patterns of the three antennas within the array look identical, see Figure 8.17.

If we decrease the element-to-element spacing by 20 mm, the coupling values increase, see Figure 8.18, but remain low. Another decrease of 20 mm further increases the coupling values, see Figure 8.19, but the effect on the radiation patterns is marginal. Although the effect on the radiation pattern is marginal, we do see from the reflection coefficients

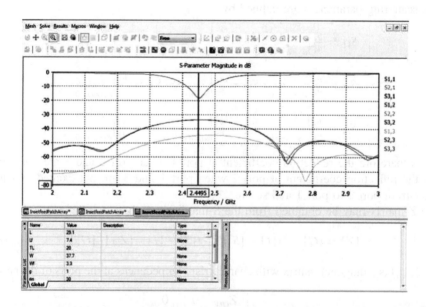

Figure 8.16 Amplitudes of the scattering parameters of the array antenna shown in Figure 8.14.

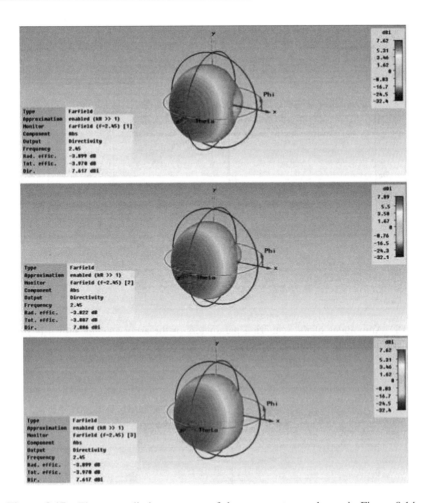

Figure 8.17 Element radiation patterns of the array antenna shown in Figure 8.14.

(S_{11}, S_{22} and S_{33}) that the input impedances of the three antennas have changed from their isolated values. The mutual coupling changes the current densities on the antennas, which is apparent immediately in the input impedance and thus the reflection coefficient. The fact that it is not that clearly visible in the radiation pattern is due to the integration of the surface current density over the antenna surface to obtain the radiated power. This integration has a smoothing out effect on the changes in surface current density.

Finally, we decrease the element-to-element spacing to 2 mm, see Figure 8.20.

The coupling between adjacent antenna elements has now reached a level of approximately −10 dB, see Figure 8.21, and the coupling effects are now visible in the radiation patterns, see Figure 8.22.

We do see that the radiation patterns of the two outward elements are symmetric with respect to the yz-plane, as was to be expected.

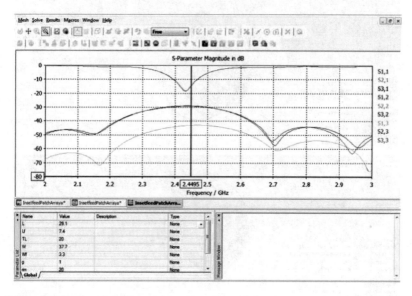

Figure 8.18 Amplitudes of the scattering parameters of the array antenna shown in Figure 8.14 with the element distance 20 mm decreased.

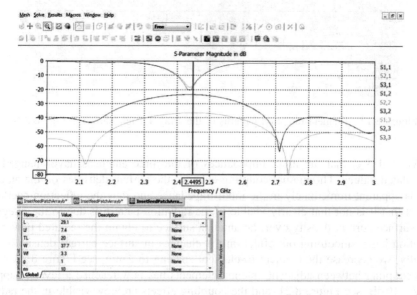

Figure 8.19 Amplitudes of the scattering parameters of the array antenna shown in Figure 8.14 with the element distance decreased by 40 mm.

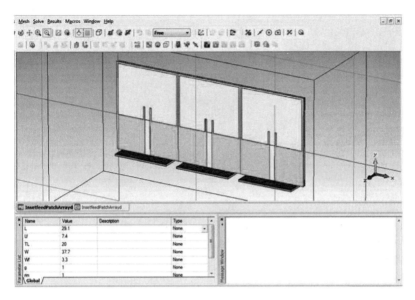

Figure 8.20 Linear array of three rectangular, inset feed, microstrip patch antennas as analyzed in CST-MWS. Dimensions are similar to those in Figure 8.14 except for a 2 mm spacing between the elements.

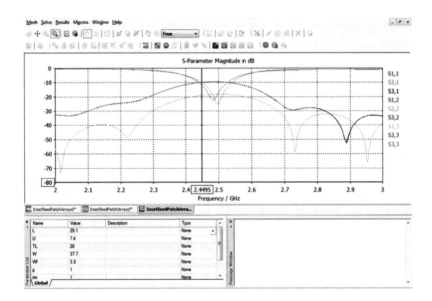

Figure 8.21 Amplitudes of the scattering parameters of the array antenna shown in Figure 8.20.

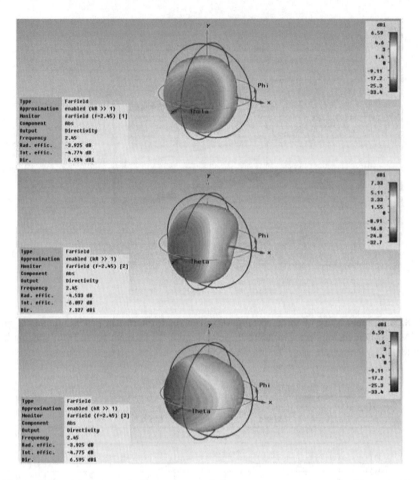

Figure 8.22 Element radiation patterns of the array antenna shown in Figure 8.20.

8.6.2 Antenna Diversity

Array antennas have found their place in mobile communications. Not only as base station antennas, which is obvious, but also inside mobile handsets. There they are employed for antenna diversity, a technique in which two or more antennas are used to improve the quality of a wireless link. This improvement is realized by combining or selecting antennas for a maximum signal strength at reception in a multipath environment. Especially in urban and indoor environments a line-of-sight between transmitter (base station) and receiver (mobile handset) is not available and the received signal has undergone many reflections along multiple paths before reaching the receiver. At each reflection, a phase shift may be introduced. Through all the phase shifts and the time delays, the signals over the multiple paths may interfere destructively at the receiver antenna location and result in deep signal fade, causing loss of the transmission link. If two or more receiving antennas are being used, each antenna will experience a different interference environment. If the receiving antennas are displaced, differently oriented or have different receive patterns, it is likely that one of the antennas, or a subset of the antennas, will experience a signal of sufficient strength.

Antenna diversity can take the form of:

Spatial diversity. By physically separating two or more antennas, multipath components having different phases will be incident on the antennas. The distance between the antennas is in the order of half a wavelength.

Polarization diversity. The employed receiver antennas are orthogonally polarized. The multipath components will, in general, have undergone polarization changes upon traveling through different media.

Radiation pattern diversity. Here the different directivities of the receive antennas will take care of selecting different multipath components.

A more thorough description of antenna diversity may be found in [6].

8.6.3 Sequential Rotation and Phasing

In Chapter 7 we saw how a circularly polarized microstrip patch antenna may be designed. Implementing this circularly polarized element into an array poses several challenges. First of all, the dual, orthogonal and 90° phase delayed feeding takes up a fair amount of space and may hinder the positioning of elements at distances that avoid the creation of grating lobes. Creating the array feeding network in microstrip technology has the benefit of a simple realization technique, but the drawback is of taking up space in between the elements, as we will see in Section 8.7 for a linearly polarized array of microstrip patch antennas. Furthermore, the feeding network is prone to coupling effects giving another reason to keep the feeding network as uncomplicated as possible.

To overcome these drawbacks, we may create circular polarization by employing linearly polarized elements in an array configuration, where we rotate the elements and excite them with a 90° phase shift. This technique, which basically consists of distributing the dual, orthogonal and 90° phase delayed fed single element over an array, is known as *sequential rotation and phasing* [7, 8].

We start the explanation of sequential rotation and phasing by using two linearly polarized elements, placed a distance d apart on the x-axis of a Cartesian coordinate system, see Figure 8.23 [1].

In the figure, the linear polarization directions are indicated by arrows, and the phasing is given by the parameter ψ. In the yz-plane a circularly polarized element is created by virtue of a perpendicular projection of the two linearly polarized elements. In the xz-plane however, the phase difference starts to deviate from the required 90° for angles ϑ starting to increase from broadside. This deviation is due to the additional phase delay $\Delta\psi = k_0 d \sin(\vartheta)$ for observation directions other than broadside, as shown in Figure 8.24.

To partly overcome this drawback, we may expand the array into two directions, as shown in Figure 8.25 [1]. We deliberately use the term partly since although the circular polarization is now improved in both principal planes, in the diagonal planes the polarization deteriorates quickly when moving away from broadside ($\vartheta = 0$).

In this book and thus also in the next example we will limit ourselves to linear array antennas.

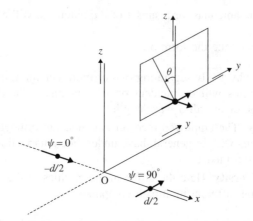

Figure 8.23 Array antenna consisting of two linearly polarized elements, phased for a 90° phase difference.

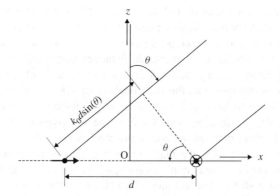

Figure 8.24 Additional spatial phase delay in the xz-plane of the two-element array antenna shown in Figure 8.23.

Example

We will construct a linear, sequentially rotated and phased array antenna consisting of two elements, as depicted in Figure 8.23. For the linearly polarized element we will use the dual polarized microstrip antenna of Figure 7.26 where we will leave one port open. We will excite the elements with $100\,\Omega$ microstrip transmission lines. These lines will be added, one lagging 90 degrees in phase (i.e. having an additional line length of a quarter of a wavelength, see Appendix F), combining to $50\,\Omega$. With this as a starting point, the required configuration is obtained within a few iterations in CST Microwave Studio®, compensating for coupling effects between transmission lines and microstrip patches. The structure is shown in Figure 8.26.

The reflection as a function of frequency is shown in Figure 8.27.

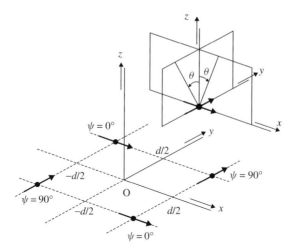

Figure 8.25 Array antenna consisting of four linearly polarized elements, phased for a 90° phase difference in the principal planes. In the xz-plane and in the yz-plane a circularly polarized element is created by virtue of perpendicular projection.

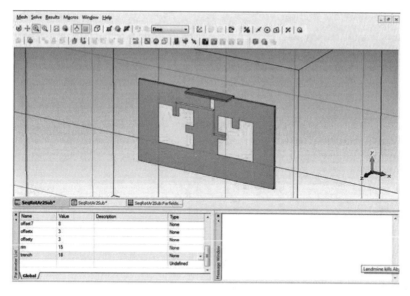

Figure 8.26 Linear array of two linearly polarized, sequentially rotated and phased microstrip patch elements.

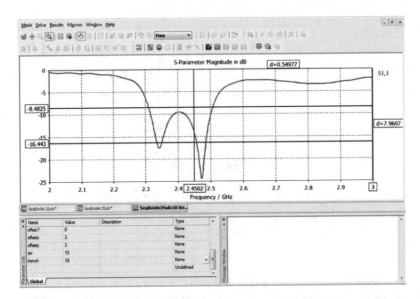

Figure 8.27 Reflection as a function of frequency for the antenna shown in Figure 8.26.

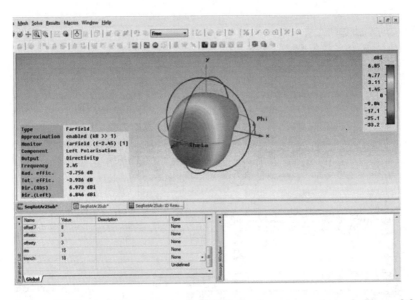

Figure 8.28 LHCP radiation pattern at 2.45 GHz for the antenna shown in Figure 8.26.

The left-hand circularly polarized (LHCP) radiation pattern is shown in Figure 8.28, the right-hand circularly polarized (RHCP) radiation patterns is shown in Figure 8.29. Both patterns are evaluated at 2.45 GHz.

The radiation patterns clearly show that the array antenna, consisting of linearly polarized elements, creates an LHCP radiation pattern. The figures also clearly demonstrate that

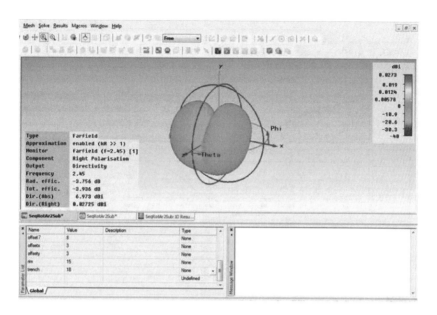

Figure 8.29 RHCP radiation pattern at 2.45 GHz for the antenna shown in Figure 8.26.

the LHCP signal is dominant in the yz-plane and that the RHCP signal stays low in the plane that is perpendicular to the linear array axis. In the xz-plane, the cross polarization comes up rather fast when changing the aspect angle ϑ from perpendicular to grazing.

This is shown in more detail in the radiation pattern cuts for the yz-plane ($\varphi = \frac{\pi}{2}$) in Figures 8.30 and 8.31 and in the radiation pattern cuts for the xz-plane ($\varphi = 0$) in Figures 8.32 and 8.33.

These two figures show that the LHCP radiation remains larger than the RHCP signal for increasing angle ϑ in the yz-plane, that is the plane perpendicular to the linear array axis. In this plane, no additional phase shift is added to the phase difference introduced between the two elements. In the xz-plane, that is in the plane containing the linear array antenna, the situation is different.

These Figures show that in the plane containing the array, moving away from broadside adds an additional phase shift between the two elements. As a result, the cross polarized (RHCP) signal increases in strength rapidly for increasing angle ϑ.

8.7 Array Antenna Design

For a certain application we want an antenna, operating at 2.45 GHz, having a gain that is about four times that of a single microstrip patch antenna. The antenna beam needs to be small in the horizontal direction and broad in the vertical direction. The input impedance needs to be 50 Ω. The antenna needs to be realized as an electrically conducting pattern on a grounded FR4 slab.

From the specifications given above, it is clear that the antenna may be realized as a (horizontal) four-element linear array of microstrip patches. The patches will be fed by

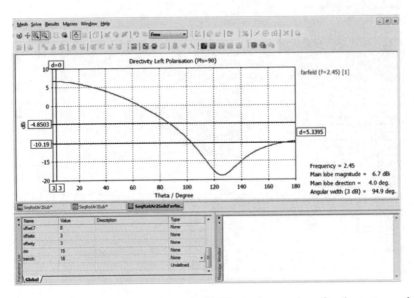

Figure 8.30 LHCP radiation pattern cut at 2.45 GHz in the yz-plane for the antenna shown in Figure 8.26.

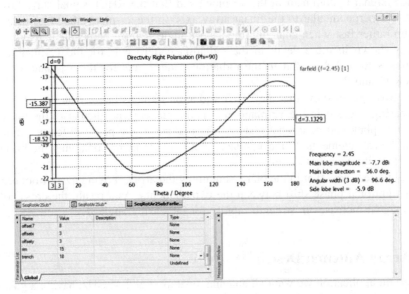

Figure 8.31 RHCP radiation pattern cut at 2.45 GHz in the yz-plane for the antenna shown in Figure 8.26.

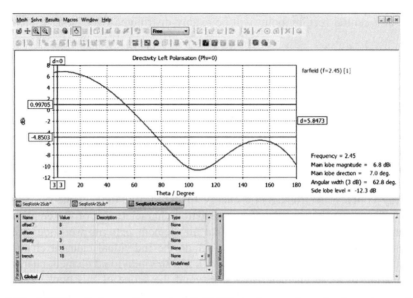

Figure 8.32 LHCP radiation pattern cut at 2.45 GHz in the xz-plane for the antenna shown in Figure 8.26.

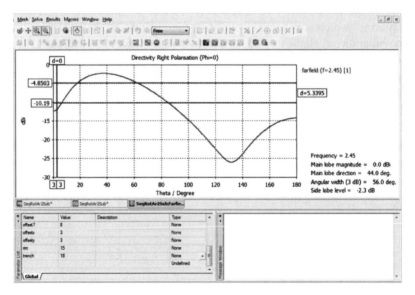

Figure 8.33 RHCP radiation pattern cut at 2.45 GHz in the xz-plane for the antenna shown in Figure 8.26.

microstrip transmission lines and these lines will be combined in such a way that the power delivered at a central feeding point will be evenly distributed over the four elements.

8.7.1 Theory

From the theory described in this chapter we know that, in order to avoid the occurrence of grating lobes, the elements should not be placed further apart than one free-space wavelength. In fact, we want the microstrip patch elements to be placed much closer than a wavelength, so that we do not even 'see' the slope of the grating lobe appear in the radiation pattern. So the width of the microstrip patch element should also be less than one free space wavelength. Since the elements will thus be closely spaced, the *mutual coupling* between the elements will not be negligible. Each radiating element in the array couples electromagnetic energy to the surrounding elements. This *mutual coupling* will change the impedance and radiation characteristics of the elements. In the theory developed thus far, we assumed that all element patterns were identical, which now turns out to be not true.[7] We could include the individual element characteristics, $S_{ei}(\vartheta)$, in equation (8.5), but then we need to know the individual element characteristics. The calculation of these characteristics is beyond the scope of this book. Therefore, in the design of the array antenna we will make use of CST-MWS. We start by designing a microstrip patch radiating element as we did before, that is we apply the aperture theory to create an initial design and then fine-tune this design using the full-wave analysis software. Then, we combine two of these elements in a sub-array, which we will fine-tune for the frequency of 2.45 GHz and finally we will combine two of these sub-arrays in the final array, which we – again – will fine-tune using CST-MWS.

The combination of elements and subarrays will be accomplished using a so-called *corporate* feeding structure, see Figure 8.34.

The length of the microstrip patch antenna element, L, will be calculated as half the wavelength in the FR4 substrate:

$$L = \frac{1}{2} \frac{\lambda_0}{\sqrt{\varepsilon_r}}, \tag{8.51}$$

where λ_0 is the wavelength in free space and ε_r is the relative permittivity of the substrate.

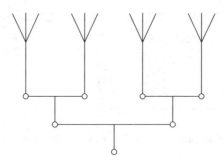

Figure 8.34 Four-element linear array with corporate feeding network.

---

[7] At least, they are not identical if the elements are closely spaced. For wider spacings the mutual coupling effect becomes less severe.

The feeding network will be realized in microstrip transmission line technology. When the characteristic impedance, Z_0, of the transmission line is known, as well as the height, d, of the substrate, the width, W, of the line can be calculated, using [2]

$$
\frac{W}{d} = \begin{cases} \dfrac{8e^A}{e^{2A}-2} & \text{for } \dfrac{W}{d} < 2, \\[2ex] \dfrac{2}{\pi}\left[B - 1 - \ln(2B - 1) + \dfrac{\varepsilon_r - 1}{2\varepsilon_r}\left\{\ln(B - 1) + 0.39 - \dfrac{0.61}{\varepsilon_r}\right\}\right] & \text{for } \dfrac{W}{d} > 2. \end{cases}
\tag{8.52}
$$

where

$$
A = \frac{Z_0}{60}\sqrt{\frac{\varepsilon_r + 1}{2}} + \frac{\varepsilon_r - 1}{\varepsilon_r + 1}\left(0.23 + \frac{0.11}{\varepsilon_r}\right),
\tag{8.53}
$$

and

$$
B = \frac{377\pi}{2Z_0\sqrt{\varepsilon_r}}.
\tag{8.54}
$$

Although we want our array antenna to have a 'standard' $50\,\Omega$ input impedance,[8] this does not mean that all our microstrip transmission line sections in the feeding network will have a $50\,\Omega$ characteristic impedance. At every T-splitter in the feeding network, see Figure 8.34, the characteristic impedance of the outgoing lines – and hence the impedance of the attached loads – should be twice the value of the characteristic impedance of the input line. Then, the parallel impedances at the output equal the impedance at the input and impedance matching is achieved.

If, however, we start at the input of the array antenna with $50\,\Omega$, we end up with transmission lines having a characteristic impedance of $200\,\Omega$. Equations (8.52)–(8.54) then learn that for a 1.6 mm thick FR4 substrate with $\varepsilon_r = 4.28$, the line width becomes 48.5 μm. This value is difficult to realize with standard photo-etching techniques. Therefore we choose to design the microstrip patch radiators to have an input impedance of $100\,\Omega$ and only use microstrip transmission lines having a characteristic impedance of $50\,\Omega$ or $100\,\Omega$. The line widths we have to deal with are then 3.1 mm and 0.84 mm, respectively. Within the feed network, impedance transformers – converting $50\,\Omega$ to $100\,\Omega$ and vice versa – then need to be applied. The details of these transformers will be discussed in the following section.

8.7.2 A Linear Microstrip Patch Array Antenna

We start with the design of a single microstrip patch radiator. Then we will combine two of these into a two-element sub-array and finally we will combine two of these sub-arrays to form the final array.

---

[8] Most microwave hardware is specified for an input or output impedance of $50\,\Omega$. This specific value is a round-off compromise between the optimum impedance value for power handling ($30\,\Omega$) and the optimum impedance value for low loss ($77\,\Omega$) in air-dielectric coaxial cables [3].

8.7.2.1 Microstrip Patch Radiator

We will use an edge-feed microstrip antenna, see Figure 8.35(a). Although a microstrip inset feed, as shown in Figure 8.35(b) would lead to a more compact radiator, we choose the edge-feed for ease of construction.

The length of the microstrip radiator will be tuned for resonance at a frequency of 2.45 GHz. The width of the antenna will be used to tune the input impedance.

We start with an antenna having a length following from equation (8.32): $L = 28$ mm. We take the identical width and use a microstrip transmission line, having a width of 0.84 mm. With the use of CST-MWS we first tune the width of the transmission line, by looking at the port impedance and then, in a few iterations, we tune the length and width of the microstrip patch antenna for an impedance match to 100 Ω at a frequency of 2.45 GHz. The length of the patch then becomes 28 mm, the width of the patch becomes 52 mm and the width of the transmission line becomes 0.80 mm. Figure 8.36 shows a screenshot of the radiating element and Figure 8.37 shows the reflection coefficient (in dB) as a function of frequency, relative to 100 Ω:

$$|S_{11}| \, (\text{dB}) = 20 \log \left| \frac{Z_{Ant} - 100}{Z_{Ant} + 100} \right| (\text{dB}). \tag{8.55}$$

We will consider reflection levels below -10 dB as acceptable. Figure 8.37 then shows that the antenna is acceptable for frequencies ranging (roughly) from 2.4 to 2.45 GHz. The radiation pattern at 2.45 GHz is shown in Figure 8.38.

The radiation pattern shows that neither the microstrip transmission line nor the finite-sized ground plane causes disturbances in the radiation pattern. The radiation efficiency is about 52% . The reason for this rather low efficiency is the use of lossy FR4 for substrate instead of using a dedicated microwave laminate.

8.7.2.2 Two-Element Sub-Array

We will now combine two single microstrip patch radiators into a two-element sub-array. There, two identical radiators are placed next to each other. The placing is not too close, to minimize mutual coupling effects but is also restricted to a maximum distance of

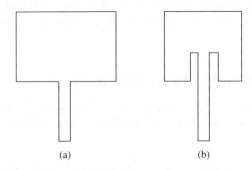

(a)   (b)

Figure 8.35   Rectangular microstrip patch antenna top view. (a) Edge microstrip feed. (b) Inset microstrip feed.

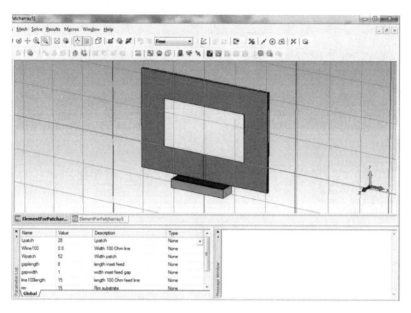

Figure 8.36 Layout of a single, rectangular microstrip patch antenna, edge-excited with a 100 Ω microstrip transmission line.

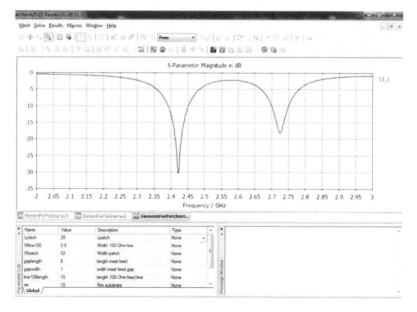

Figure 8.37 Reflection coefficient of a single microstrip patch antenna as a function of frequency.

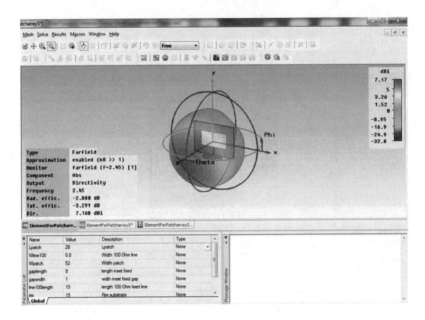

Figure 8.38 Radiation pattern of a single microstrip patch antenna at 2.45 GHz.

one free space wavelength, see equation (8.19). Crossing this one free space wavelength border would result in the creation of grating lobes in the radiation pattern. The 100 Ω transmission lines of the individual radiators are combined to form (together with the 100 Ω radiators) a 50 Ω load to the 50 Ω transmission line, connected to the junction of the 100 Ω transmission lines. As shown in [4], 90 degree angles in the transmission lines will not give rise to serious radiation, so 90 degree angles have been applied in the feeding network of the two-element sub-array without applying miters, that is cutting the corners. The two-element sub-array, obtained after a few iterations in CST-MWS, is shown in Figure 8.39. In the first iterations, the 50 Ω transmission line width has been tuned to a value of 3.3 mm. In the subsequent iterations, a center-to-center distance between the patches of 30.4 mm has been obtained.

Figure 8.40 shows the reflection coefficient (in dB), relative to 50 Ω, as a function of frequency:

$$|S_{11}| \, (\mathrm{dB}) = 20 \log \left| \frac{Z_{Ant} - 50}{Z_{Ant} + 50} \right| (\mathrm{dB}). \tag{8.56}$$

The figure shows that the mutual coupling has worked to our advantage and has broadened the −10 dB bandwidth with respect to a single radiator. The radiation pattern of the two-element sub-array is shown in Figure 8.41.

As expected, the radiated beam has become narrower in the horizontal direction. The gain enhancement is less than a factor two. This is due to the losses in the FR4 substrate.

Before we combine two of these sub-arrays into the final array, we need to discuss the design of an impedance transformer with which we will transform the 50 Ω input impedance of the two-element sub-array to 100 Ω.

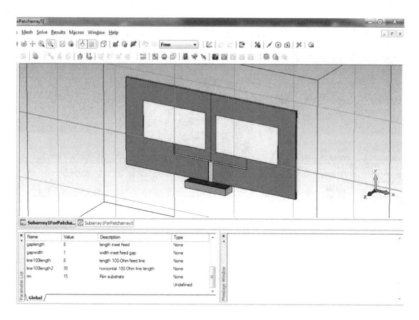

Figure 8.39   Layout of a two-element sub-array with 50 Ω to two times 100 Ω feeding network.

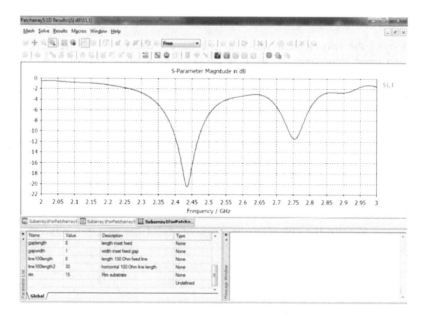

Figure 8.40   Reflection coefficient of a two-element sub-array as a function of frequency.

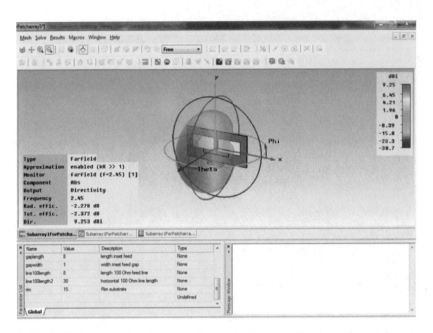

Figure 8.41 Radiation pattern of a two-element sub-array of microstrip patch antennas at 2.45 GHz.

8.7.2.3 Microstrip Impedance Transformer

At the input of the two-element sub-array we have created – in the frequency band of interest – an input impedance of $50\,\Omega$. If we were to combine two of these sub-arrays, we would end up, at the input of the T-splitter, with an input impedance of $25\,\Omega$. As explained before, we want to restrict ourselves to microstrip transmission lines of $50\,\Omega$ and $100\,\Omega$ only. Therefore we need to employ a so-called impedance transformer, see Figure 8.42.

This impedance transformer, known as the *quarter lambda transformer*, consists of a transmission line of length $\frac{\lambda_g}{4}$, where λ_g is the wavelength in the medium of the transmission line, having a characteristic impedance Z_0 that is given by [5]

$$Z_0 = \sqrt{Z_{01} Z_{02}}, \tag{8.57}$$

where Z_{01} and Z_{02} are the characteristic impedances of the two transmission lines that need to be impedance matched.

8.7.2.4 Final Array

The splitter/combiner network joining the two two-element sub-arrays, realized in microstrip technology, will have a top view as shown in Figure 8.43.

With the shown splitter/combiner, the line lengths in $50\,\Omega$ and $100\,\Omega$ can be freely chosen. Only the lengths of the $\sqrt{100 \cdot 50} = 71\,\Omega$ transmission line sections need to be kept constant. This makes the creation of the final array very easy. We maintain the

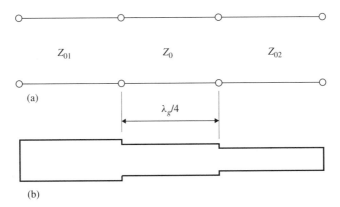

Figure 8.42 (Quarter lambda) impedance transformer. (a) Schematic view. (b) Top view of a microstrip quarter wavelength transformer.

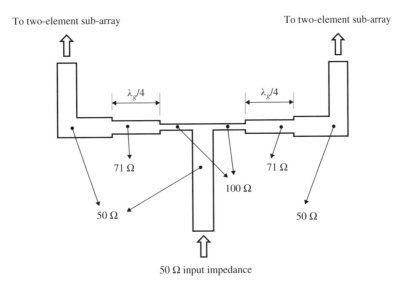

Figure 8.43 Top view of a microstrip $50-100\,\Omega$ splitter/combiner circuit.

element spacing by positioning the second two-element sub-array and get a layout as shown in Figure 8.44. The transformer section length and width have been tuned to 20 mm and 1.6 mm, respectively.

In Figure 8.45 we show the reflection coefficient as a function of frequency, and the radiation pattern of the array antenna is shown in Figure 8.46.

With respect to the two-element sub-array, we see that the $-10\,\mathrm{dB}$ frequency bandwidth has decreased a bit. This must be due to the narrowband behavior of the quarter lambda impedance transformer. The beam of the array antenna has further narrowed with respect to the two-element sub-array. The gain increase, however, is less than a factor two, due to the losses in the FR4 substrate.

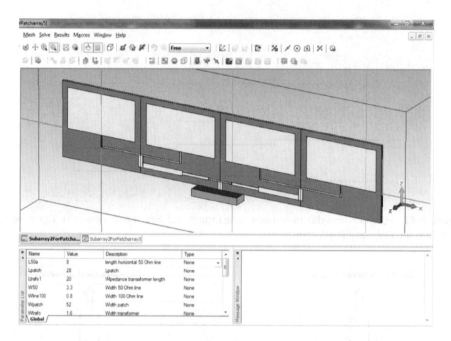

Figure 8.44 Layout of the final four-element, 50 Ω microstrip patch array antenna.

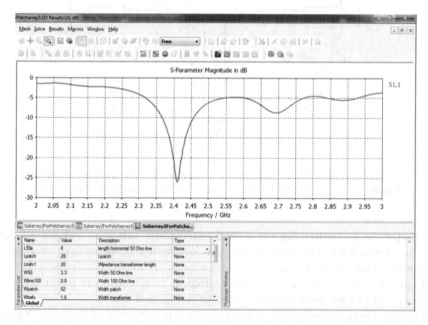

Figure 8.45 Reflection coefficient of a four-element microstrip patch array antenna as a function of frequency.

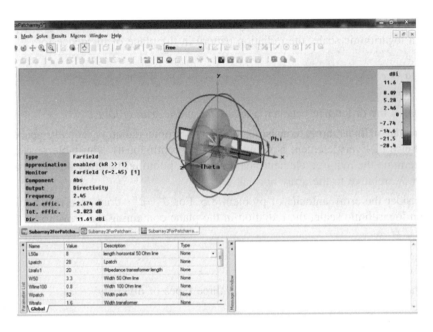

Figure 8.46   Radiation pattern of a four-element microstrip patch array antenna at 2.45 GHz.

8.8   Problems

8.1 Consider a linear array antenna consisting of four isotropic radiators, equally spaced by a distance d. A plane wave is incident on the array at an angle ϑ with respect to the array normal, in the plane of the array.

(a) What are the path lengths the plane wave has to travel to the different, elements with respect to the first element encountered by the wave?

(b) What are the phases of the signals received by the elements relative to that same element?

8.2 Consider the array of problem 8.1. Draw, in a rectangular plot and on a logarithmic scale, the radiation pattern cut in the plane of the array for

(a) $d = \frac{\lambda_0}{2}$.

(b) $d = \lambda_0$.

(c) $d = \frac{3\lambda_0}{2}$.

8.3 Consider the array of problem 8.1 and replace the isotropic radiators by half- wave dipole antennas, positioned perpendicular to the array axis and parallel to the plane of incidence (i.e. z-directed dipoles for a linear array on the x-axis the xz-plane being the plane of incidence). Draw, in a rectangular plot and on a logarithmic scale, the radiation pattern in the plane of incidence for $d = \lambda_0$.

8.4 Consider the array of problem 8.1 and replace the isotropic radiators by dipole antennas, slightly shorter than half a wavelength, positioned parallel to the array axis (i.e.

x-directed dipoles for a linear array on the x-axis). Draw, in a rectangular plot and on a logarithmic scale, the radiation pattern in the plane of incidence for

(a) $d = \frac{\lambda_0}{2}$.

(b) $d = \lambda_0$.

(c) $d = \frac{3\lambda_0}{2}$.

8.5 What is the difference between a *side lobe* and a *grating lobe*?

8.6 Consider a linear array consisting of four isotropic radiators, equally spaced by a distance d. Every element is equipped with a continuously adjustable phase shifter. Determine the required element phase shifts for pointing the array beam to the direction ϑ_0, relative to the array normal.

8.7 Consider the array antenna of problem 8.6. For $\vartheta_0 = \frac{\pi}{6}$, draw in a rectangular plot, on a logarithmic scale, the radiation in the plane containing the array for

(a) $d = \frac{\lambda_0}{2}$.

(b) $d = \lambda_0$.

(c) $d = \frac{3\lambda_0}{2}$.

8.8 What is diversity? Name and describe three types of diversity.

References

1. Hubregt J. Visser, *Array and Phased Array Antenna Basics*, John Wiley & Sons, Chichester, UK, 2005.
2. David M. Pozar, *Microwave Engineering, second edition*, John Wiley & Sons, New York, 1989.
3. 'Why 50 Ohms?', *Microwave Encyclopedia*, available at: http://www.microwaves101.com/encyclopedia/why50ohms.cmf.
4. Hubregt J. Visser, 'Equivalent Length Design Equations for Right-Angled Microstrip Bends', *Proceedings EuCAP2007*, 6 pp., 2007.
5. Robert E. Collin, *Foundations for Microwave Engineering*, McGraw-Hill, New York, 1992.
6. Yi Huang and Kevin Boyle, *Antennas from Theory to Practice*, John Wiley & Sons, Chichester, UK, 2008.
7. T. Teshirogi, M. Tanaka and W. Chujo, 'Wideband Circularly Polarized Array Antenna with Sequential Rotation and Phase Shift of Elements', *Proceedings International Symposium Antennas & Propagation*, Japan, pp. 117–120, August 1985.
8. John Huang, 'A Technique for an Array to Generate Circular Polarization with Linearly Polarized Elements', *IEEE Transactions on Antennas and Propagation*, Vol. AP-34, No.9, pp. 1113–1124, September 1986.

Appendix A

Effective Aperture and Directivity[1]

The effective aperture of an antenna is uniquely related to its directivity. By using the directivity and effective aperture of a short dipole, which are relatively easily calculated, a general interrelation between effective aperture and directivity, valid for any antenna, will be derived.

Consider the two-antenna communication system of Figure A.1. System *1* can be the transmitter, while system *2* is the receiver, or the other way around. The antennas are displaced a distance *r* apart and are assumed to be lined up with respect to polarization and directivity. The directivity of antenna *1* is D_T, its maximum effective aperture is A_{emT}. Directivity and maximum effective area of antenna *2* are, respectively, D_R and A_{emR}. We start by considering the first option: antenna *1* is transmitting and antenna *2* is receiving.

The total radiated power by antenna *1* is P_T. If antenna *1* were an isotropic radiator, the power density, S_0, at distance *r* from antenna *1* would be

$$S_0 = \frac{P_T}{4\pi r^2}. \tag{A.1}$$

Due to the directive properties of antenna *1*, the actual power density, S_T at distance *r* is

$$S_T = S_0 D_T = \frac{P_T D_T}{4\pi r^2}. \tag{A.2}$$

The power received by antenna *2*, P_R is then

$$P_R = S_T A_{emR} = \frac{P_T D_T A_{meR}}{4\pi r^2}, \tag{A.3}$$

[1] Portions of text in this Appendix have been reproduced from: Visser, H. 'Array and Phased Array Antenna Basics'. This includes the following text: The effective (p. 231) ... of an antenna (p. 233). Reproduced with permissions from John Wiley & Sons, Ltd.

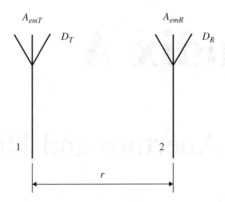

Figure A.1 Two-antenna communication system. The antennas are displaced a distance r and are assumed to be lined up with respect to polarization and directivity. The directivity of antenna *1* is D_T, its effective aperture is A_{emT}. Directivity and effective area of antenna *2* are, respectively, D_R and A_{emR}.

where A_{emR} is the maximum effective aperture of antenna *2*. Rearranging this equation gives

$$D_T A_{emR} = \frac{P_R}{P_T} \left(4\pi r^2\right).$$ (A.4)

If we now let antenna *2* transmit P_T and we look at the received power at antenna *1*, which, by virtue of reciprocity, is equal to P_R, we find

$$D_R A_{emT} = \frac{P_R}{P_T} \left(4\pi r^2\right),$$ (A.5)

so

$$\frac{D_T}{A_{emT}} = \frac{D_R}{A_{emR}}.$$ (A.6)

If we now assume that in the two-antenna system, the transmitting antenna is an isotropic radiator, then $D_T = 1$ and the above equation transforms into

$$A_{emISO} = \frac{A_{emR}}{D_R},$$ (A.7)

which means that

> the maximum effective aperture of an isotropic radiator is equal to the ratio of effective aperture and directivity of any antenna.

If we now take, for example, a short dipole, we may relatively easily calculate the effective area and directivity as, see Section 2.4.1,

$$A_{em} = \frac{3}{8\pi}\lambda^2, \tag{A.8}$$

$$D = \frac{3}{2}, \tag{A.9}$$

where λ is the used wavelength. Therefore the maximum effective area of an isotropic radiator is

$$A_{em_{ISO}} = \frac{\lambda^2}{4\pi}, \tag{A.10}$$

and thus *for any antenna*

$$\frac{\lambda^2}{4\pi} = \frac{A_{em}}{D}. \tag{A.11}$$

This gives us the sought-after relation between maximum effective aperture, A_{em}, and directivity, D, of an antenna

$$D = \frac{4\pi A_{em}}{\lambda^2}. \tag{A.12}$$

Appendix B

Vector Formulas

Although the material provided in Chapter 3 should be sufficient to manipulate any vector formula that will be encountered in antenna theory, it will be convenient to have a list of most commonly used vector formulas. A selection of these is presented in this appendix.

In the following we will make use of the vectors $\mathbf{A}$, $\mathbf{B}$ and $\mathbf{C}$ and the scalars Φ and Ψ.

We start by writing the gradient, divergence and curl operations in rectangular coordinates:

$$grad\Phi = \nabla\Phi = \hat{\mathbf{u}}_x \frac{\partial\Phi}{\partial x} + \hat{\mathbf{u}}_y \frac{\partial\Phi}{\partial y} + \hat{\mathbf{u}}_z \frac{\partial\Phi}{\partial z}, \tag{B.1}$$

$$div\mathbf{A} = \nabla \cdot \mathbf{A} = \frac{\partial A_x}{\partial x} + \frac{\partial A_y}{\partial y} + \frac{\partial A_z}{\partial z}, \tag{B.2}$$

$$curl\mathbf{A} = \nabla \times \mathbf{A}$$

$$= \hat{\mathbf{u}}_x \left(\frac{\partial A_z}{\partial y} - \frac{\partial A_y}{\partial z} \right) + \hat{\mathbf{u}}_y \left(\frac{\partial A_x}{\partial z} - \frac{\partial A_z}{\partial x} \right) + \hat{\mathbf{u}}_z \left(\frac{\partial A_y}{\partial x} - \frac{\partial A_x}{\partial y} \right). \tag{B.3}$$

Also

$$\nabla^2\Phi = \frac{\partial^2\Phi}{\partial x^2} + \frac{\partial^2\Phi}{\partial y^2} + \frac{\partial^2\Phi}{\partial z^2}, \tag{B.4}$$

$$\nabla^2\mathbf{A} = \hat{\mathbf{u}}_x \nabla^2 A_x + \hat{\mathbf{u}}_y \nabla^2 A_y + \hat{\mathbf{u}}_z \nabla^2 A_z. \tag{B.5}$$

A list of useful vector identities is given below:

$$\nabla(\Phi + \Psi) = \nabla\Phi + \nabla\Psi, \tag{B.6}$$

$$\nabla \cdot (\mathbf{A} + \mathbf{B}) = \nabla \cdot \mathbf{A} + \nabla \cdot \mathbf{B}, \tag{B.7}$$

$$\nabla \times (\mathbf{A} + \mathbf{B}) = \nabla \times \mathbf{A} + \nabla \times \mathbf{B}. \tag{B.8}$$

Antenna Theory and Applications, First Edition. Hubregt J. Visser.
© 2012 John Wiley & Sons, Ltd. Published 2012 by John Wiley & Sons, Ltd.

$$\nabla \left(\Phi \Psi \right) = \Phi \nabla \Psi + \Psi \nabla \Phi, \tag{B.9}$$

$$\nabla \cdot \left(\Phi \mathbf{A} \right) = \mathbf{A} \cdot \nabla \Phi + \Phi \nabla \cdot \mathbf{A}, \tag{B.10}$$

$$\nabla \times \left(\Phi \mathbf{A} \right) = \nabla \Phi \times \mathbf{A} + \Phi \nabla \times \mathbf{A}. \tag{B.11}$$

$$\nabla \cdot \left(\mathbf{A} \times \mathbf{B} \right) = \mathbf{B} \cdot \nabla \times \mathbf{A} - \mathbf{A} \cdot \nabla \times \mathbf{B}, \tag{B.12}$$

$$\nabla \times \left(\mathbf{A} \times \mathbf{B} \right) = \mathbf{A} \nabla \cdot \mathbf{B} - \mathbf{B} \nabla \cdot \mathbf{A} + \left(\mathbf{B} \cdot \nabla \right) \mathbf{A} - \left(\mathbf{A} \cdot \nabla \right) \mathbf{B}, \tag{B.13}$$

$$\nabla \left(\mathbf{A} \cdot \mathbf{B} \right) = \left(\mathbf{A} \cdot \nabla \right) \mathbf{B} + \left(\mathbf{B} \cdot \nabla \right) \mathbf{A} + \mathbf{A} \times \left(\nabla \times \mathbf{B} \right) + \mathbf{B} \times \left(\nabla \times \mathbf{A} \right). \tag{B.14}$$

$$\nabla \cdot \nabla \Phi = \nabla^2 \Phi, \tag{B.15}$$

$$\nabla \cdot \nabla \times \mathbf{A} = 0, \tag{B.16}$$

$$\nabla \times \nabla \Phi = 0, \tag{B.17}$$

$$\nabla \times \nabla \times \mathbf{A} = \nabla \left(\nabla \cdot \mathbf{A} \right) - \nabla^2 \mathbf{A}. \tag{B.18}$$

$$\mathbf{A} \cdot \mathbf{B} \times \mathbf{C} = \mathbf{B} \cdot \mathbf{C} \times \mathbf{A} = \mathbf{C} \cdot \mathbf{A} \times \mathbf{B}, \tag{B.19}$$

$$\mathbf{A} \times \left(\mathbf{B} \times \mathbf{C} \right) = \mathbf{B} \left(\mathbf{A} \cdot \mathbf{C} \right) - \mathbf{C} \left(\mathbf{A} \cdot \mathbf{B} \right). \tag{B.20}$$

Appendix C

Complex Analysis

The use of complex variables in solving problems in the applied sciences, as in electrical engineering, appears to be a very valuable tool. Especially when dealing with sinusoidal excitations, the introduction of complex variables will simplify the solution process. Before this simplification is fully appreciated though, we have to deal first with the somewhat awkward concept of complex numbers.

C.1 Complex Numbers

We are all familiar with the real numbers and the permitted and non-permitted operations on real numbers. So, it is, for example, permitted to calculate the square root of the number 3.79 ($\sqrt{3.79} = 1.95$), but the square root of -4 does not exist.

The *complex numbers* allow for the last square root to exist, through the introduction of so-called *imaginary numbers* alongside the real numbers. Any complex number consists of a real part and an imaginary part and is generally denoted as

$$c = a + jb, \tag{C.1}$$

where c is a complex number, a is the real part of the complex number

$$a = \Re(c), \tag{C.2}$$

and b is the imaginary part of the complex number

$$b = \Im(c). \tag{C.3}$$

j is the imaginary unit[1] that exhibits the special characteristic

$$j^2 = -1. \tag{C.4}$$

---

[1] Mathematicians and physicists use the symbol i for the imaginary unit, but since in electrical engineering this symbol is already reserved for current, electrical engineers use the symbol j instead.

Antenna Theory and Applications, First Edition. Hubregt J. Visser.
© 2012 John Wiley & Sons, Ltd. Published 2012 by John Wiley & Sons, Ltd.

Before we move on to the arithmetic concerning complex numbers, we first define the *complex conjugate* of a complex number.

The complex conjugate of a complex number c, denoted c^*, is defined by replacing j by $-j$ everywhere in the complex number. So, if

$$c = a + jb, \tag{C.5}$$

then

$$c^* = a - jb, \tag{C.6}$$

The addition and subtraction of complex numbers is straightforward. If $c = a + jb$ and $d = e + jf$, then

$$c + d = a + jb + e + jf = (a + e) + j(b + f), \tag{C.7}$$

$$c - d = a + jb - (e + jf) = (a - e) + j(b - f). \tag{C.8}$$

Multiplication makes use of the special characteristic of the imaginary unit

$$e \cdot (a + jb) = ea + jeb, \tag{C.9}$$

$$(e + jf) \cdot (a + jb) =$$

$$ea + jeb + jaf + j^2 fb =$$

$$(ea - fb) + j(eb + af). \tag{C.10}$$

Multiplication of a complex number with its complex conjugate results in

$$cc^* = (a + jb)(a - jb)$$

$$= a^2 + jab - jab - j^2 b^2$$

$$= a^2 + b^2$$

$$= |c|^2. \tag{C.11}$$

For division, use is made of the complex conjugate. If $c = a + jb$ and $d = e + jf$, then

$$\frac{c}{d} = \frac{cd^*}{dd^*}$$

$$= \frac{(a + jb)(e - jf)}{e^2 + f^2}$$

$$= \frac{ae + bf}{e^2 + f^2} + j\frac{be - af}{e^2 + f^2} \tag{C.12}$$

Complex numbers may be graphically represented in the *complex plane*, see Figure C.1. The real part of a complex number is plotted along the horizontal axis, and the imaginary part of the complex number, multiplied by j, is plotted along the vertical axis.

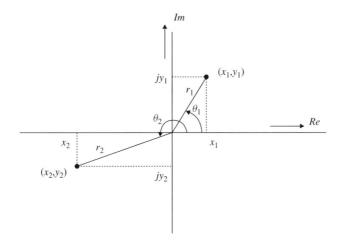

Figure C.1   Complex numbers (x_1, y_1) and (x_2, y_2) in the complex plane.

Another way of representing complex numbers instead of using Cartesian coordinates is using polar coordinates. For $z_1 = x_1 + jy_1$, see Figure C.1, the polar form is

$$z_1 = r_1 \left[\cos(\vartheta_1) + j\sin(\vartheta_1)\right], \tag{C.13}$$

where

$$r = |z_1|, \tag{C.14}$$

and

$$\vartheta_1 = \arctan\left(\frac{\Im(z_1)}{\Re(z_1)}\right). \tag{C.15}$$

If we differentiate the expression of z_1 to ϑ_1, we get

$$\begin{aligned}
\frac{dz_1}{d\vartheta_1} &= r_1 \left[-\sin(\vartheta_1) + j\cos(\vartheta_1)\right] \\
&= jr_1 \left[\cos(\vartheta_1) + j\sin(\vartheta_1)\right] \\
&= jz_1,
\end{aligned} \tag{C.16}$$

and therefore

$$\begin{aligned}
z_1 &= r_1 \left[\cos(\vartheta_1) + j\sin(\vartheta_1)\right] \\
&= r_1 e^{j\vartheta_1}.
\end{aligned} \tag{C.17}$$

As a final check, we verify that

$$z_1^* = r_1 e^{-j\vartheta_1}, \tag{C.18}$$

and therefore

$$z^* z = r_1^2. \tag{C.19}$$

C.2   Use of Complex Variables

Let us first look at an electric capacitor, C. The current 'through' a capacitor, i, is described as function of the voltage, v, across the capacitor

$$i = C\frac{dv}{dt}.\tag{C.20}$$

Now, assume that we are dealing with sinusoidal signals (currents and voltages)

$$v = V_1\cos(\omega t),\tag{C.21}$$

where $\omega = 2\pi f$ is the angular frequency.

The current 'through' the capacitor is now given by

$$\begin{aligned}
i &= C\frac{d}{dt}[V_1\cos(\omega t)]\\
&= -V_1 C\omega\sin(\omega t).
\end{aligned}\tag{C.22}$$

Next, we look at an inductor, L. The voltage across the inductor, v, is described as function of the current, i, through the inductor

$$v = L\frac{di}{dt}.\tag{C.23}$$

For sinusoidal signals, $i = I_1\cos(\omega t)$,

$$\begin{aligned}
v &= L\frac{d}{dt}[I_1\cos(\omega t)]\\
&= -I_1 L\omega\sin(\omega t).
\end{aligned}\tag{C.24}$$

If we work with complex signals, we get for the capacitor

$$v = \Re\left(V_1 e^{j\omega t}\right),\tag{C.25}$$

$$\begin{aligned}
i &= C\Re\left(\frac{d}{dt}V_1 e^{j\omega t}\right)\\
&= C\Re\left(j\omega V_1 e^{j\omega t}\right)\\
&= -V_1 C\omega\sin(\omega t),
\end{aligned}\tag{C.26}$$

and for the inductor

$$i = \Re\left(I_1 e^{j\omega t}\right),\tag{C.27}$$

$$\begin{aligned}
v &= L\Re\left(\frac{d}{dt}I_1 e^{j\omega t}\right)\\
&= L\Re\left(j\omega I_1 e^{j\omega t}\right)\\
&= -I_1 L\omega\sin(\omega t),
\end{aligned}\tag{C.28}$$

We see that for the complex signals the operator $\frac{d}{dt}$ is replaced by the multiplication by $j\omega$.

Therefore, we can define a complex impedance for the capacitor as

$$
\begin{aligned}
Z_C &= \frac{V_1 e^{j\omega t}}{jC\omega V_1 e^{j\omega t}} \\
&= \frac{1}{j\omega C}.
\end{aligned}
\tag{C.29}
$$

Similarly, we may define a complex impedance for the inductor as

$$
\begin{aligned}
Z_L &= \frac{jL\omega I_1 e^{j\omega t}}{I_1 e^{j\omega t}} \\
&= j\omega L.
\end{aligned}
\tag{C.30}
$$

To show the ease of working with complex signals, we will calculate the input impedance of a parallel circuit consisting of a resistor, R, an inductor, L, and a capacitor, C, see Figure C.2.

The complex input admittance Y_{in} is

$$
\begin{aligned}
Y_{in} &= \frac{1}{R} + \frac{1}{j\omega L} + j\omega C \\
&= \frac{1}{R} - j\frac{1}{\omega L}\left(1 - \omega^2 LC\right),
\end{aligned}
\tag{C.31}
$$

and the complex input impedance is therefore found to be

$$
\begin{aligned}
Z_{in} &= \frac{1}{Y_{in}} \\
&= \frac{1}{\frac{1}{R} - j\frac{1}{\omega L}\left(1 - \omega^2 LC\right)} \\
&= \frac{R}{1 - j\frac{R}{\omega L}\left(1 - \omega^2 LC\right)} \\
&= \frac{R\left[1 + j\frac{R}{\omega L}\left(1 - \omega^2 LC\right)\right]}{1 + \frac{R^2}{\omega^2 L^2}\left(1 - \omega^2 LC\right)^2}.
\end{aligned}
\tag{C.32}
$$

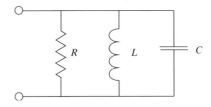

Figure C.2 Parallel electric circuit consisting of resistor R, inductor L and capacitor C.

Finally, the 'real-world' input resistance is found as the real part of the complex input impedance

$$R_{in} = \Re\{Z_{in}\}$$

$$= \frac{R}{1 + \frac{R^2}{\omega^2 L^2}\left(1 - \omega^2 LC\right)^2}. \tag{C.33}$$

The phase is found to be, see also Figure C.2

$$\Psi_{R_{in}} = \arctan\left(\frac{\Im\{Z_{in}\}}{\Re\{Z_{in}\}}\right)$$

$$= \frac{1}{\omega L}\left(1 - \omega^2 LC\right). \tag{C.34}$$

We see that the calculation of the input resistance of the parallel circuit, using complex numbers, is really straightforward, while a direct solution based on time-derivatives of real sinusoidal currents and voltages would have been far more complicated.

So, the general idea in dealing with a sinusoidal excitation, S_{in} (voltage or current) is to create a complex number, S_{in}^c, such that $\Re\left\{S_{in}^c\right\} = S_{in}$ (or $\Im\left\{S_{in}^c\right\} = S_{in}$), then calculate the response of the complex signal, S_{out}^c – which is easy since time-derivatives are replaced by multiplications by $j\omega$ – and finally extract the real response, S_{out} by taking $S_{out} = \Re\left\{S_{out}^c\right\}$ (or $S_{out} = \Im\left\{S_{out}^c\right\}$).

Appendix D

Physical Constants and Material Parameters

Throughout the text we have used several physical constants, such as the velocity of light in free space. In the examples we have been using non-perfect electrically conducting metals such as copper and have used dielectric materials. For convenience, in this appendix we have grouped the physical constants used in this book and have provided a list of material properties frequently encountered.

The physical constants used frequently within this book are

- permittivity of free space: $\varepsilon_0 = 8.854 \cdot 10^{-12}\,\mathrm{Fm}^{-1}$
- permeability of free space: $\mu_0 = 4\pi \cdot 10^{-7}\,\mathrm{Hm}^{-1}$
- velocity of light in free space: $c_0 = 2.998 \cdot 10^8\,\mathrm{ms}^{-1}$
- impedance of free space: $\eta_0 = 376.7 \approx 120\pi\,\Omega$

The electrical conductivities of some commonly used metals and liquids at room temperature are given in Table D.1.

Table D.1 Electrical conductivities of some common metals and fluids

Material	Conductivity Sm^{-1}
aluminum	$3.816 \cdot 10^7$
copper	$5.813 \cdot 10^7$
gold	$4.098 \cdot 10^7$
iron	$1.03 \cdot 10^7$
lead	$4.56 \cdot 10^6$
nickel	$1.449 \cdot 10^7$
stainless steel	$1.1 \cdot 10^6$
solder	$7.0 \cdot 10^6$
distilled water	$2 \cdot 10^{-4}$
sea water	4

Antenna Theory and Applications, First Edition. Hubregt J. Visser.
© 2012 John Wiley & Sons, Ltd. Published 2012 by John Wiley & Sons, Ltd.

Table D.2 Dielectric constants and loss tangents of some materials

Material	Relative Permittivity ε_r	Loss Tangent $\tan\delta$
FR4 (3 GHz)	4.28	0.016
glass Pyrex (3 GHz)	4.82	0.0054
plexiglas (3 GHz)	2.60	0.0057
polyethylene (10 GHz)	2.25	0.0004
polystyrene (10 GHz)	2.54	0.00033
silicon (10 GHz)	11.9	0.004
styrofoam 103.7 (3 GHz)	1.03	0.0001
Teflon (10 GHz)	2.08	0.0004
water distilled (3 GHz)	76.7	0.157

The dielectric constants and loss tangents of some materials are given in Table D.2 [1–3].

References

1. David M. Pozar, *Microwave Engineering, second edition*, John Wiley & Sons, New York, 1998.
2. Simon Ramo, John R. Whinnery and Theodore van Duzer, *Fields and Waves in Communication Electronics, second edition*, John Wiley & Sons, New York, 1984.
3. Richard C. Johnson, *Antenna Engineering Handbook, third edition*, McGraw Hill, New York, 1993.

Appendix E

Two-Port Network Parameters

A microwave network with two terminals may be analyzed by considering the network as an equivalent two-port. The signals at the terminals or ports of the two-port may be described in terms of voltages and currents and the interaction between these signals in terms of impedances or admittances. Alternatively, the signals may be described in terms of incident and reflected waves and the interaction in terms of scattering.

An arbitrary two-port may be described in several ways. Here we will discuss the two-port in terms of impedance parameters (Z), admittance parameters (Y), chain or ABCD parameters and scattering parameters (S).

For the description in terms of Z, Y or ABCD parameters we refer to Figure E.1(a). For the description in terms of S parameters we refer to Figure E.1(b).

The relation between the voltages and currents in terms of impedance parameters is given by

$$\begin{pmatrix} V_1 \\ V_2 \end{pmatrix} = \begin{pmatrix} Z_{11} & Z_{12} \\ Z_{21} & Z_{22} \end{pmatrix} \begin{pmatrix} I_1 \\ I_2 \end{pmatrix}. \tag{E.1}$$

The relation between the currents and voltages in terms of admittance parameters is given by

$$\begin{pmatrix} I_1 \\ I_2 \end{pmatrix} = \begin{pmatrix} Y_{11} & Y_{12} \\ Y_{21} & Y_{22} \end{pmatrix} \begin{pmatrix} V_1 \\ V_2 \end{pmatrix}. \tag{E.2}$$

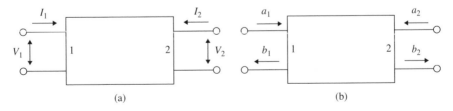

Figure E.1 Two-port definition. (a) Using voltages and currents. (b) Using incident and reflected waves.

Antenna Theory and Applications, First Edition. Hubregt J. Visser.
© 2012 John Wiley & Sons, Ltd. Published 2012 by John Wiley & Sons, Ltd.

Table E.1 Two-port parameter conversion for voltage- and current-based matrices

	Z		Y		ABCD	
Z	Z_{11} Z_{12}		$\frac{Y_{22}}{\|Y\|}$ $-\frac{Y_{12}}{\|Y\|}$		$\frac{A}{C}$ $\frac{AD-BC}{C}$	
	Z_{21} Z_{22}		$-\frac{Y_{21}}{\|Y\|}$ $\frac{Y_{11}}{\|Y\|}$		$\frac{1}{C}$ $\frac{D}{C}$	
Y	$\frac{Z_{22}}{\|Z\|}$ $-\frac{Z_{12}}{\|Z\|}$		Y_{11} Y_{12}		$\frac{D}{B}$ $\frac{BC-AD}{B}$	
	$-\frac{Z_{21}}{\|Z\|}$ $\frac{Z_{11}}{\|Z\|}$		Y_{21} Y_{22}		$-\frac{1}{B}$ $\frac{A}{B}$	
ABCD	$\frac{Z_{11}}{Z_{21}}$ $\frac{\|Z\|}{Z_{21}}$		$-\frac{Y_{22}}{Y_{21}}$ $-\frac{1}{Y_{21}}$		A B	
	$\frac{1}{Z_{21}}$ $\frac{Z_{22}}{Z_{21}}$		$-\frac{\|Y\|}{Y_{21}}$ $-\frac{Y_{11}}{Y_{21}}$		C D	

Table E.2 Two-port parameter conversion for scattering and voltage- and current-based matrices

	S		Z	
S	S_{11} S_{12}		$\frac{(Z_{11}-Z_0)(Z_{22}+Z_0)-Z_{12}Z_{21}}{\Delta Z}$	$\frac{2Z_{12}Z_0}{\Delta Z}$
	S_{21} S_{22}		$\frac{2Z_{21}Z_0}{\Delta Z}$	$\frac{(Z_{11}+Z_0)(Z_{22}-Z_0)-Z_{12}Z_{21}}{\Delta Z}$

	S		Y	
S	S_{11} S_{12}		$\frac{(Y_0-Y_{11})(Y_0+Y_{22})+Y_{12}Y_{21}}{\Delta Y}$	$-\frac{2Y_{12}Y_0}{\Delta Y}$
	S_{21} S_{22}		$-\frac{2Y_{21}Y_0}{\Delta Y}$	$\frac{(Y_0+Y_{11})(Y_0-Y_{22})+Y_{12}Y_{21}}{\Delta Y}$

	S		ABCD	
S	S_{11} S_{12}		$\frac{A+\frac{B}{Z_0}-CZ_0-D}{A+\frac{B}{Z_0}+CZ_0+D}$	$\frac{2(AD-BC)}{A+\frac{B}{Z_0}+CZ_0+D}$
	S_{21} S_{22}		$\frac{2}{A+\frac{B}{Z_0}+CZ_0+D}$	$\frac{-A+\frac{B}{Z_0}-CZ_0+D}{A+\frac{B}{Z_0}+CZ_0+D}$

The relation between the voltages and currents in terms of chain parameters is given by

$$\begin{pmatrix} V_1 \\ I_1 \end{pmatrix} = \begin{pmatrix} A & B \\ C & D \end{pmatrix} \begin{pmatrix} V_2 \\ -I_2 \end{pmatrix}. \tag{E.3}$$

Chain matrices are used when several two-ports are cascaded. The overall chain matrix is obtained by multiplying the chain matrices of the individual two-ports.

Table E.3 Two-port parameter conversion for scattering and voltage- and current-based matrices

	S	
S	S_{11} S_{12}	
	S_{21} S_{22}	
Z	$Z_0 \dfrac{(1+S_{11})(1-S_{22})+S_{12}S_{21}}{(1-S_{11})(1-S_{22})-S_{12}S_{21}}$	$Z_0 \dfrac{2S_{12}}{(1-S_{11})(1-S_{22})-S_{12}S_{21}}$
	$Z_0 \dfrac{2S_{21}}{(1-S_{11})(1-S_{22})-S_{12}S_{21}}$	$Z_0 \dfrac{(1-S_{11})(1+S_{22})+S_{12}S_{21}}{(1-S_{11})(1-S_{22})-S_{12}S_{21}}$
Y	$Y_0 \dfrac{(1-S_{11})(1+S_{22})+S_{12}S_{21}}{(1+S_{11})(1+S_{22})-S_{12}S_{21}}$	$-Y_0 \dfrac{2S_{12}}{(1+S_{11})(1+S_{22})-S_{12}S_{21}}$
	$-Y_0 \dfrac{2S_{21}}{(1+S_{11})(1+S_{22})-S_{12}S_{21}}$	$Y_0 \dfrac{(1+S_{11})(1-S_{22})+S_{12}S_{21}}{(1+S_{11})(1+S_{22})-S_{12}S_{21}}$
ABCD	$\dfrac{(1+S_{11})(1-S_{22})+S_{12}S_{21}}{2S_{21}}$	$Z_0 \dfrac{(1+S_{11})(1+S_{22})-S_{12}S_{21}}{2S_{21}}$
	$\dfrac{1}{Z_0}\dfrac{(1-S_{11})(1-S_{22})-S_{12}S_{21}}{2S_{21}}$	$\dfrac{(1-S_{11})(1+S_{22})+S_{12}S_{21}}{2S_{21}}$

The relation between the incident and reflected waves is given by

$$\begin{pmatrix} b_1 \\ b_2 \end{pmatrix} = \begin{pmatrix} S_{11} & S_{12} \\ S_{21} & S_{22} \end{pmatrix} \begin{pmatrix} a_1 \\ a_2 \end{pmatrix}. \tag{E.4}$$

The Z, Y, ABCD and S matrices may be converted into each other. The interrelations between the voltage- and current-based matrices are given in Table E.1.

The interrelations between the scattering and voltage- and current-based matrices are given in Tables E.2 and E.3.

Appendix F

Transmission Line Theory

At microwave frequencies, the wavelengths have become so small that the physical dimensions of transmission lines and even those of lumped elements, such as resistors, capacitors and inductors, are in the order of these wavelengths. This means that at these frequencies we have to consider effects of waves, such as standing waves and reflections. Depending on the type of transmission line under consideration, these effects may be best characterized by employing a field description or employing a circuit description. In this appendix we will limit ourselves to a circuit description of transmission lines.

The microwave frequency range is somewhat arbitrary, but in practice, frequencies between 300 MHz and 30 GHz may be considered as being in the microwave spectrum.

F.1 Distributed Parameters

A general long (i.e. long with respect to wavelength) two-wire transmission line may be characterized by distributed transmission line parameters, see Figure F.1.

Here R is the sum of resistances in both conductors per unit of length, G is the conductivity per unit of length, L is the self-inductance per unit of length and C is the capacitance per unit of length.

When the distributed transmission line parameters are known, the characteristic impedance, Z_0, and propagation constant, γ_0, may be calculated as

$$Z_0 = \sqrt{\frac{R + j\omega L}{G + j\omega C}}, \tag{F.1}$$

$$\gamma_0 = \sqrt{(R + j\omega L)(G + j\omega C)}. \tag{F.2}$$

We will demonstrate this, as well as state the definitions for characteristic impedance and propagation constant, using an infinitesimal length, Δz, of transmission line as shown in Figure F.2.

Antenna Theory and Applications, First Edition. Hubregt J. Visser.
© 2012 John Wiley & Sons, Ltd. Published 2012 by John Wiley & Sons, Ltd.

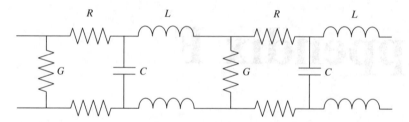

Figure F.1   Distributed transmission line parameters in a general long two-wire transmission line.

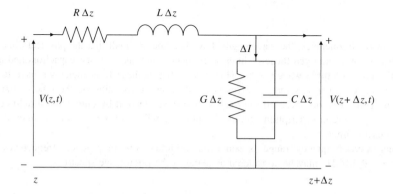

Figure F.2   Equivalent network for an infinitesimal length of transmission line.

Applying the Kirchhoff *voltage* law to the circuit of Figure F.2 gives

$$V(z, t) = R\Delta z I(z, t) + L\Delta z \frac{\partial I(z, t)}{\partial t} + V(z + \Delta z, t), \tag{F.3}$$

or, after rearranging terms,

$$-\frac{V(z + \Delta z, t) - V(z, t)}{\Delta z} = R I(z, t) + L\frac{\partial I(z, t)}{\partial t}. \tag{F.4}$$

In the limit $\Delta z \rightarrow 0$, this equation reduces to

$$-\frac{\partial V(z, t)}{\partial z} = R I(z, t) + L\frac{\partial I(z, t)}{\partial t}. \tag{F.5}$$

Next, applying the Kirchhoff *current* law to the circuit of Figure F.2 gives

$$I(z, t) = I(z + \Delta z, t) + \Delta I$$

$$= I(z + \Delta z, t) + G\Delta z V(z + \Delta z, t) + C\Delta z \frac{\partial V(z + \Delta z, t)}{\partial t}, \tag{F.6}$$

which may be written as

$$-\frac{I(z + \Delta z, t) - I(z, t)}{\Delta z} = G V(z + \Delta z, t) + C\frac{\partial V(z + \Delta z, t)}{\partial t}. \tag{F.7}$$

In the limit $\Delta z \to 0$, this equation reduces to

$$-\frac{\partial I(z,t)}{\partial z} = GV(z,t) + C\frac{\partial V(z,t)}{\partial t}. \tag{F.8}$$

If we now suppose time harmonic signals, that is signals having a (co)sinusoidal time-dependency, we may describe the voltages and currents using complex quantities:

$$V(z,t) = \Re\left\{V_s(z)e^{j\omega t}\right\}, \tag{F.9}$$

$$I(z,t) = \Re\left\{I_s(z)e^{j\omega t}\right\}, \tag{F.10}$$

where $\Re\{x\}$ means the real part of complex argument x. The parameter $\omega = 2\pi f$ is the angular frequency, where f is the frequency.

Substitution of equations (F.9) and (F.10) into equations (F.5) and (F.8) gives

$$-\frac{dV_s}{dz} = (R + j\omega L)I_s, \tag{F.11}$$

$$-\frac{dI_s}{dz} = (G + j\omega C)V_s. \tag{F.12}$$

Taking the derivative to z of equation (F.11) and substituting equation (F.12) into that equation yields

$$\frac{d^2 V_s}{dz^2} = (R + j\omega L)(G + j\omega C)V_s. \tag{F.13}$$

This equation may be written as

$$\frac{d^2 V_s}{dz^2} - \gamma^2 V_s = 0, \tag{F.14}$$

where

$$\gamma = \alpha + j\beta = \sqrt{(R + j\omega L)(G + j\omega C)}. \tag{F.15}$$

Equation (F.14) is known as the *wave equation* or *Helmholtz equation*, γ is known as the *propagation constant*. The propagation constant consists of an *attenuation constant*, α, and a *phase constant*, β, where $\beta = \frac{2\pi}{\lambda}$, λ being the wavelength.

For the current, a wave equation of identical form may be derived

$$\frac{d^2 I_s}{dz^2} - \gamma^2 I_s = 0. \tag{F.16}$$

Solutions of the Helmholtz equations for voltage and current are

$$V_s(z) = V_0^+ e^{-\gamma z} + V_0^- e^{+\gamma z}, \tag{F.17}$$

$$I_s(z) = I_0^+ e^{-\gamma z} + I_0^- e^{+\gamma z}, \tag{F.18}$$

where V_0^+ and I_0^+ are the amplitudes of, respectively, voltage and current waves traveling in the positive z-direction and V_0^- and I_0^- are the amplitudes of, respectively, voltage and current waves traveling in the negative z-direction.

The characteristic impedance, Z_0, of the transmission line is defined as the ratio of, in the positive direction, traveling voltage and current:

$$Z_0 = \frac{V_0^+}{I_0^+}. \tag{F.19}$$

By use of equations (F.11), (F.12), (F.17) and (F.18), we find for the characteristic impedance

$$Z_0 = \frac{V_0^+}{I_0^+} = \frac{R + j\omega L}{\gamma} = \frac{\gamma}{G + j\omega C}, \tag{F.20}$$

and upon substitution of equation (F.15)

$$Z_0 = \sqrt{\frac{R + j\omega L}{G + j\omega C}}. \tag{F.21}$$

F.2   Guided Waves

We have seen in the previous section that by introducing complex quantities, we may describe voltage and current at any place on a transmission line as a superposition of a (voltage or current) wave traveling in the positive direction and one traveling in the negative direction. We used this concept, in combination with that of distributed transmission line parameters, to derive expressions for the propagation constant and characteristic impedance of a transmission line.

In this section we will further explore this guided wave[1] property of transmission lines to derive practical parameters such as *voltage standing wave ratio* (VSWR), *reflection factor*, *characteristic impedance* and *input impedance*. These parameters prove to be very useful in designing microwave networks, as they allow us to design subsystems and predict the behavior of the interconnected subsystems on the basis of the values of these parameters.

In order to analyze transmission lines, we will take the general two-wire transmission line of Figure F.3 and look in detail at the voltage between and the current through the two wires.

Voltage and current in the long transmission line can propagate as a wave going in the positive s-direction and as a wave going in the negative s-direction. For the wave propagating in the positive s-direction

$$v^+(s, t) = \Re\left\{V^+(s)e^{j\omega t}\right\}, \tag{F.22}$$

where

$$V^+(s) = Ae^{-\gamma s}, \tag{F.23}$$

and

$$i^+(s, t) = \Re\left\{I^+(s)e^{j\omega t}\right\}, \tag{F.24}$$

---

[1] *Guided* waves as opposed to *unguided* waves which we encounter in a radio link.

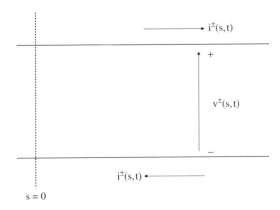

Figure F.3 Voltage and current of a two-wire transmission line.

where

$$I^+(s) = \frac{A}{Z_0}e^{-\gamma s}, \tag{F.25}$$

Z_0 being the characteristic impedance of the transmission line. A is, for the moment, an unknown complex amplitude coefficient.

For the wave propagating in the negative s-direction

$$v^-(s, t) = \Re\left\{V^-(s)e^{j\omega t}\right\}, \tag{F.26}$$

where

$$V^-(s) = Be^{+\gamma s}, \tag{F.27}$$

and

$$i^-(s, t) = \Re\left\{I^-(s)e^{j\omega t}\right\}, \tag{F.28}$$

where

$$I^-(s) = -\frac{B}{Z_0}e^{+\gamma s}. \tag{F.29}$$

B is, for the moment, an unknown complex amplitude coefficient.

The propagation constant is also in general complex,

$$\gamma = \alpha + j\beta, \tag{F.30}$$

where α is the attenuation constant and β is the phase constant. When the transmission line is lossless, $\alpha = 0$ and the amplitude of the wave is constant over the transmission line. In that situation $R = G = 0$ and equation (F.21) reveals that for that situation the characteristic impedance is real. Using the sign convention of Figure F.3 makes the characteristic impedance positive.

The phase constant β is related to the wavelength λ. Whenever s increases by an amount equal to λ, the same phase must be encountered: $\beta\lambda = 2\pi$ and thus

$$\beta = \frac{2\pi}{\lambda}. \tag{F.31}$$

β is also known as the *wave number*.

F.2.1 VSWR and Reflection Factor

When a traveling wave at some point is totally or partially reflected, a standing wave is created. The ratio of the absolute values of the complex voltage wave amplitude at maximum and at minimum is known as the *voltage standing wave ratio* (VSWR), *S*.

$$S = \frac{|V|_{max}}{|V|_{min}} = \frac{|V^+| + |V^-|}{|V^+| - |V^-|}. \tag{F.32}$$

The reflection factor, ρ, is defined as

$$\rho = \frac{V^-}{V^+} = \frac{B}{A} e^{2\gamma s}. \tag{F.33}$$

From equations (F.32) and (F.33) follows

$$S = \frac{1 + |\rho|}{1 - |\rho|}, \tag{F.34}$$

and

$$|\rho| = \frac{S - 1}{S + 1}. \tag{F.35}$$

F.2.2 Impedance and Relative Impedance

The impedance, Z, which is a function of the position s along the transmission line, just like ρ, is defined as

$$Z = \frac{V}{I} = \frac{V^+ + V^-}{I^+ + I^-}. \tag{F.36}$$

The relative impedance, z, is the impedance Z normalized to the characteristic impedance of the transmission line

$$z = \frac{Z}{Z_0} = \frac{1 + \frac{B}{A} e^{2\gamma s}}{1 - \frac{B}{A} e^{2\gamma s}}. \tag{F.37}$$

In determining the impedance we suppose the transmission line tp be cut at position s. The impedance is related to the part of the transmission line to the right of the cut. An excitation voltage $V(s)$ then results in a current $I(s)$.

Substitution of equation (F.33) into equation (F.37) leads to

$$z = \frac{1 + \rho}{1 - \rho}, \tag{F.38}$$

and

$$\rho = \frac{z - 1}{z + 1}. \tag{F.39}$$

F.3   Input Impedance of a Transmission Line

Assume a transmission line of length l, meaning $0 \leq s \leq l$. We will add a subscript i to the parameters Z, z, S and ρ when they refer to the input of the transmission line ($s = 0$). When they refer to the output of the transmission line ($s = l$), we will use a subscript u. For an arbitrary position s ($0 < s < l$) we will use no subscript.

We now want to express the input impedance, Z_i, as function of the load impedance, Z_u, at the end of the transmission line.

By use of equation (F.33) we find

$$\rho_i = \rho_u e^{-2\gamma l}. \tag{F.40}$$

Upon substitution of this result into equation (F.38) and using equation (F.39), we find for the normalized input impedance

$$z_i = \frac{1 + \rho_u e^{-2\gamma l}}{1 - \rho_u e^{-2\gamma l}} = \frac{1 + \frac{z_u - 1}{z_u + 1} e^{-2\gamma l}}{1 - \frac{z_u - 1}{z_u + 1} e^{-2\gamma l}}. \tag{F.41}$$

After some straightforward, though lengthy calculations,[2] this equation may be rewritten into

$$z_i = \frac{z_u \cosh(\gamma l) + \sinh(\gamma l)}{z_u \sinh(\gamma l) + \cosh(\gamma l)}. \tag{F.42}$$

For a lossless transmission line, $\gamma = j\beta = j\frac{2\pi}{\lambda}$, so that

$$z_i = \frac{z_u + j \tan\left(2\pi \frac{l}{\lambda}\right)}{1 + j z_u \tan\left(2\pi \frac{l}{\lambda}\right)}. \tag{F.43}$$

F.4   Terminated Lossless Transmission Line

Using equation (F.43), that relates the normalized input impedance to the normalized load impedance of a lossless transmission line of length l, we will now look into some special situations.

F.4.1   Matched Load

A transmission line is terminated into a *matched load* if the load impedance is the complex conjugate of the characteristic impedance of the transmission line. Since, by virtue of equation (F.43), we assume our transmission line to be lossless, we have seen that the characteristic impedance is real. Therefore, the line is terminated into a matched load if the load impedance is equal to the characteristic impedance of the transmission line.

---

[2] Start by multiplying numerator and denominator of the equation by $\frac{1}{2}(z_u + 1) e^{\gamma l}$.

So, we have the situation that $Z_u = Z_0$ and therefore $z_u = 1$. This means that everywhere on the transmission line $Z = Z_0$ and $\rho = 0$ (no reflection). Also, $S = 1$.

F.4.2   Short Circuit

When we terminate the transmission line into a short circuit, $Z_u = 0$ and therefore also $z_u = 0$. Substitution of $z_u = 0$ into equation (F.43) gives

$$z_i = j \tan\left(\frac{2\pi l}{\lambda}\right). \tag{F.44}$$

Upon a closer inspection of this equation we see that, going from the short circuit over the transmission line to the input, we alternately, at intervals of a quarter wavelength, encounter an impedance that is either purely inductive or purely capacitive, see also Figure F.4.

A practical application may be found in the creation of capacitors and inductors by means of pieces of short-circuited transmission line.

F.4.3   Open Circuit

If we leave the transmission line open at the end, $Z_u = \infty$ and thus $z_u = \infty$. Substitution of $z_u = \infty$ into equation (F.43) gives

$$z_i = -j\frac{1}{\tan\left(\frac{2\pi l}{\lambda}\right)}. \tag{F.45}$$

We find an input impedance behavior over the transmission line analogous to the situation depicted in Figure F.4, but shifted by a quarter of a wavelength.

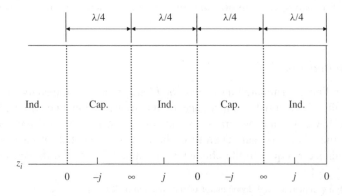

Figure F.4   Impedance behavior along a short-circuited transmission line.

F.4.4 Imaginary Unit Termination

When $Z_u = jZ_0$, remembering that Z_0 is real, $z_u = j$. Substitution of $z_u = j$ into equation (F.43) gives

$$z_i = j \tan\left(2\pi\frac{l}{\lambda} + \frac{\pi}{4}\right). \tag{F.46}$$

Again, we find an input impedance behavior similar to the situation as depicted in Figure F.4, but now shifted over one eight of a wavelength.

F.4.5 Real Termination

If Z_u is real, $z_u = r_u$. We may distinguish two situations:

F.4.5.1 $r_u < 1$

For this situation

$$|\rho| = \frac{1 - r_u}{1 + r_u}, \tag{F.47}$$

and

$$S = \frac{1}{r_u} > 1. \tag{F.48}$$

F.4.5.2 $r_u > 1$

For this situation

$$|\rho| = \frac{r_u - 1}{r_u + 1}, \tag{F.49}$$

and

$$S = r_u > 1. \tag{F.50}$$

F.5 Quarter Wavelength Impedance Transformer

We have seen that if the termination of a lossless transmission line – that is, the impedance connected at the end of the transmission line – is not identical to the characteristic impedance of that transmission line, reflections will occur.

Not only are these reflections unwanted due to the fact that they prohibit a complete signal transfer, they are also unwanted since they distort the quality of the signal transferred. When a *mismatch* exists not only at the end of the transmission line but also at the beginning, the generator-side part of the signal power will reach the load (at the end of the line) after a certain time delay.

To overcome the negative effects of a mismatch an impedance transformer may be placed in between transmission line or microwave circuit and load to make a reflection-free transition between transmission line or microwave circuit and load.

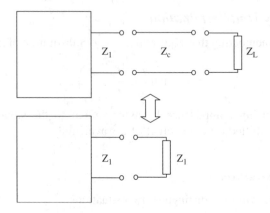

Figure F.5 Impedance matching using a quarter wavelength impedance transformer.

Such an impedance transformer may be realized very easily, employing a piece of transmission line with the right characteristic impedance.

Starting with equation (F.43), we take a piece of transmission line of characteristic impedance Z_c and length equal to a quarter of a wavelength at the frequency of operation. The load impedance is Z_L. The unnormalized input impedance is then found to be

$$Z_{in} = Z_c z_{in} = Z_c \left[\frac{Z_L + jZ_c \tan \left(\frac{\pi}{2} \right)}{Z_c + jZ_L \tan \left(\frac{\pi}{2} \right)} \right] = \frac{Z_c^2}{Z_L}. \tag{F.51}$$

So, if we take a piece of transmission line, a quarter of a wavelength long (at the operating frequency) and dimension this transmission line such that its characteristic impedance is equal to

$$Z_c = \sqrt{Z_{in} Z_L}, \tag{F.52}$$

where Z_{in} is the required input impedance (usually the impedance level of the circuit connected to the load), and place this piece of transmission line between circuit output and load, we have created a reflectionless transition from circuit to load, see Figure F.5.

The equivalent of the impedance transformer and load impedance Z_L is a new load impedance equal to Z_{in}.

Appendix G

Coplanar Waveguide (CPW)

In a coplanar waveguide (CPW) structure, metalization is present only on top of a dielectric slab. Impedance is controlled by the separation of metallic traces and not by substrate thickness. This can be advantageous, especially when making connections at high frequencies.

The cross section of a classic CPW is shown in Figure G.1. By 'classic' we mean that no ground plane is present. CPW structures on a grounded dielectric slab also exist.

The characteristic impedance, Z_0, may be calculated as [1]

$$Z_0 = \frac{30\pi}{\sqrt{\varepsilon_{eff}}} \frac{K\left(k'\right)}{K\left(k\right)}, \tag{G.1}$$

where

$$\varepsilon_{eff} = 1 + \frac{\varepsilon_r - 1}{2} \frac{K\left(k'\right) K\left(k_1\right)}{K\left(k\right) K\left(k'_1\right)}, \tag{G.2}$$

and

$$k = \frac{W}{W + 2S}, \tag{G.3}$$

$$k_1 = \frac{\sinh\left(\frac{\pi W}{4H}\right)}{\sinh\left(\frac{(W+2S)\pi}{4H}\right)}. \tag{G.4}$$

In the above, K is the complete elliptic integral of the first kind and $k' = \sqrt{\left(1 - k^2\right)}$. The ratio of complete elliptic functions in equation (G.1) may be approximated by [2]

$$\frac{K\left(k\right)}{K\left(k'\right)} \approx \begin{cases} \frac{1}{2\pi} \ln\left[2\frac{\sqrt{1+k}+\sqrt[4]{4k}}{\sqrt{1+k}-\sqrt[4]{4k}}\right] & \text{for } 1 \le \frac{K}{K'} \le \infty, \frac{1}{\sqrt{2}} \le k \le 1 \\ \frac{2\pi}{\ln\left[2\frac{\sqrt{1+k'}+\sqrt[4]{4k'}}{\sqrt{1+k'}-\sqrt[4]{4k'}}\right]} & \text{for } 0 \le \frac{K}{K'} \le 1, 0 \le k \le \frac{1}{\sqrt{2}} \end{cases}. \tag{G.5}$$

Antenna Theory and Applications, First Edition. Hubregt J. Visser.
© 2012 John Wiley & Sons, Ltd. Published 2012 by John Wiley & Sons, Ltd.

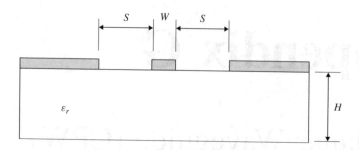

Figure G.1   Cross section of a classic Coplanar Waveguide (CPW) structure.

References

1. G. Ghione and C. Naldi, 'Analytical Formulas for Coplanar Lines in Hybrid and Monolithic Mics', Electronics Letters, Vol. 20, No. 4, pp. 179–181, February 1984.
2. W. Hilberg, 'From Approximations to Exact Relations for Characteristic Impedances', *IEEE Transactions on Microwave Theory and Techniques*, Vol. MTT-17, No. 5, pp. 259–265. May 1969.

Index

Antenna Theory and Applications, First Edition. Hubregt J. Visser.
© 2012 John Wiley & Sons, Ltd. Published 2012 by John Wiley & Sons, Ltd.